𝕴𝖓𝖙𝖊𝖑𝖑𝖎𝖌𝖊𝖓𝖙 𝕸𝖆𝖗𝖘 𝕴:

Sacred Geometry of the Mountains.

Did Da Vinci Know?

Arthur Raymond Beaubien

epiphi productions

Intelligent Mars I: Sacred Geometry of the Mountains. Did Da Vinci Know?

ISBN 9780994032102

epiphi productions
Ottawa, Ontario
Canada

www.epiphiproductions.ca

Library and Archives Canada Cataloguing in Publication

Beaubien, Arthur Raymond, 1942-, author
    Intelligent Mars I : sacred geometry of the mountains : did
Da Vinci know? / Arthur Raymond Beaubien.

Includes bibliographical references.
ISBN 978-0-9940321-0-2 (pbk.)

    1. Mars (Planet)--Geology. 2. Mars (Planet)--Surface.
3. Geometry. 4. Life on other planets. 5. Extraterrestrial
anthropology. I. Title.

QB641.B42 2015   559.9'23   C2015-900539-6

Cover design © by Arthur R. Beaubien. Mars map on cover courtesy USGS Astrogeology.

Dedication

*To all who suffer
discrimination, repression,
slavery or cruelty
at the hands of
those foolish enough
to consider themselves
superior*

About the Author

Arthur Raymond Beaubien has a daughter and lives in Ottawa, Canada. His initial degree was a B.A. in philosophy. He then pursued a career in science, earning a M.Sc. and a Ph.D. in pharmacology. He worked as a research scientist for Health Canada for over 25 years in the area of drug toxicology and is now retired. He has more than 25 scientific publications particularly in the areas of pharmacokinetics and in drug-induced hearing loss. Among his major published findings are: (1) the importance of first-pass binding to non-metabolising sites in the liver in the pharmacokinetics of the antidepressant drug imipramine (2) that the ototoxicity (destruction of hair cells in the inner ear leading to deafness) of the aminoglycoside antibiotic amikacin is a predictable function of the cumulative area under the plasma or perilymph concentration vs. time curve rather than blood levels per se (3) that amikacin has a half-life in the ototoxic compartment of about 80 days and (4) that it takes a minimum of 4 days to detect hearing loss once a toxic amount of amikacin has been administered. Although his Ph.D. degree is in pharmacology, he also has an extensive background in electrophysiology and computer programming. Upon retirement, he has devoted much of his time to writing and to his own spiritual development.

Some of his publications:

(1) *Antagonism of imipramine poisoning by anticonvulsants in the rat. Beaubien AR, Carpenter DC, Mathieu LF, MacConaill M and Hrdina PD. Toxicol. Appl. Pharmacol. 1976, 38: 1-6. (16 citations)*

(2) *Influence of dose on first-pass kinetics of 14C-imipramine in the isolated perfused rat liver. A. R. Beaubien and A. P. Pakuts. Drug Metabolism and Disposition, January 1979, 7:34-39. (16 citations)*

(3) *Incidence of amikacin ototoxicity: A sigmoid function of total drug exposure independent of plasma levels. A.R. Beaubien, S. Desjardins, E. Ormsby, A. Bayne, K. Carrier, M. J. Cauchy, R. Henri, M. Hodgen, J. Salley and A. St Pierre. Am. J. Otolaryngol. 1989, 10(4): 234-43. (22 citations)*

(4) *Delay in hearing loss following drug administration: a consistent feature of amikacin ototoxicity.* *A. R. Beaubien, S. Desjardins, E. Ormsby, A. Bayne, K. Carrier, M. J. Cauchy, R. Henri, M. Hodgen, J. Salley and A. St Pierre. Acta Oto-laryngologica, 1990, Vol. 109, No. 5-6, pp 345-352. (15 citations)*

(5) *Evidence that amikacin ototoxicity is related to total perilymph area under the concentration-time curve regardless of concentration. Beaubien AR, Ormsby E, Bayne A, Carrier K, Crossfield G, Downes M, Henri R, Hodgen M. Antimicrob Agents Chemother. 1991 June; 35(6):1070–1074. (29 citations)*

(6) *Toxicodynamics and toxicokinetics of amikacin in the guinea pig cochlea. A.R. Beaubien, K. Karpinski and E. Ormsby. Hearing Reseach, Volume 83, Issues 1-2, March 1995, pp 62-79. (10 citations)*

Acknowledgements

The author wishes to express his gratitude to Gayle Peterson and his daughter Kery for the irreplaceable caring, interactions and teaching that they have provided him over the years. He also wishes to thank George A. Neville for his invaluable enduring friendship, suggestions and support, and to Denyse Robillard, Liana Manole, Gabrielle Markvorsen, and Keith Bailey for their generous gift of time and many helpful comments in their review of this manuscript.

Cover: Tracing of Da Vinci's Vitruvian man superimposed on a map of Olympus Mons and the 3 Tharsis Montes. The full explanation of the rationale is given in the book. The map segment of Mars is courtesy of USGS Astrogeology.

Contents

Foreword

Intelligent Mars is a 3 part series devoted to the study of Martian topography. With the dawning of the era of remotely controlled spacecraft, a wealth of new data about the Martian landscape has become available which far exceeds the resolution of our best terrestrial telescopes. Much of this data has been released to the public on the Internet, allowing private citizens the opportunity to study Mars on a level unattainable by expert astronomers prior to the space age. I have availed myself of this new data to analyse the layout and shape not only of the Martian mountains and craters but of other landforms as well. The results of this analysis provide a completely different picture of Mars from that given to us by NASA. The first book of the series focuses on how the mountains are not hap-hazardously located, but are arranged in patterns according to the principles of sacred geometry. The second book studies the subject of ancient Martian coordinate systems and special-purpose craters. The third book deals with large scale artifacts and a very unusual crater which appears to have influenced the organization of much of the Martian landscape.

The subject matter of these books is so outrageous to our present view of reality that it would seem to belong more appropriately in the science fiction category rather than serious science. Undoubtedly that is where many will put it without a second thought. Others will refuse to accept any evidence of artificiality regardless of overwhelming statistical proof simply because it goes against the "authoritative" portrayal of Mars by NASA. There may even be others who will take the subject matter seriously and will want to discredit it in any manner that they can because they have a vested interest in suppressing the truth.

Once our minds have been inseminated with a set of beliefs or axioms, we are programmed to interpret all of our reality in terms of these starting points. Our beliefs and axioms seem to be stored in a part of the brain which acts like read-only memory (ROM) in a computer. Actually this type of memory is more analogous to erasable programmable read-only memory (EPROM) but it seems that it can only be altered by a major disruption in our lives (e.g., expending a considerable effort to re-educate ourselves, exposure to extreme events, repetitive presentation of another viewpoint, religious conversion). If information is encountered which is incompatible with our beliefs/axioms, the overwhelming tendency is to reject the information rather than question the starting principles. We will do almost anything to protect our familiar/comfortable belief systems (e.g., shoot the messenger, go into denial, repress the information, avoid the information, explain it away with ridiculous arguments, examine it incompletely). In the case of Mars, some of the axioms given to us are: (1) it has always been devoid of intelligent life and (2) that its topography is strictly the result of random physical phenomena such as volcanic eruptions which create mountains and meteoric bombardments which create craters.

We have been living for many centuries with the notion that our planet hosts the only life forms in the universe, or at least, the only intelligent life form. With the modern era of space travel, we have opened ourselves up a little to the possibility of the existence of extraterrestrial life, and have engaged in several SETI (search for extraterrestrial intelligence) projects and the encoding of ourselves in 1977 on the Pioneer 10 spacecraft. A good number of us even believe in UFO's and are hoping that contact will someday be achieved.

Despite the very substantial efforts of governments to discredit all UFO sightings, crop circles, ancient high tech megalithic structures on earth, and observations of artifacts on Mars and the moon, there remains a hard core of individuals who risk ridicule by publicly professing their belief in the existence of intelligent beings beyond our planet. I hope that this book will not only reach these individuals, but also encourage many others to break out of the fish-pond mentality that limits us so much.

The implications of this series of books extend far beyond simple acknowledgement of extraterrestrial intelligence. If we accept the conclusion that Mars has hosted intelligent life hundreds of millions and even billions of years ago, we will have to re-examine everything, from religion to science, to ancient history, to our relationship with the rest of the solar system, galaxy, and the universe - even the likely existence of other dimensional realities and universes. All of our normal reference points have now changed forever, and we must chaotically struggle for a time to establish new ones which will ultimately serve us much better. The process is scary, but one we must go through if we hope to survive as a species in the new reality of a shared universe with others. On the other hand it is also fascinating and can be awe-inspiring. Who would have ever thought it possible that very substantial progress would ever be within our reach on some of the most frustrating questions we humans have had about the phenomenon of our very existence? We have that opportunity now. Let's make use of it!

Arthur R. Beaubien

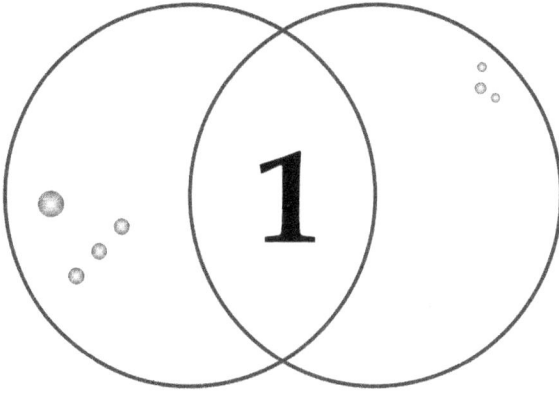

Inspiration

H ave you ever experienced a situation where you don't see an item that you are looking for even though it's in plain view? You don't see it because you didn't expect it to be in the spot where it actually was. Psychologists would attribute this phenomenon to "violated expectations" or "inattentional blindness". In a classical experiment by J.J. Bruner and Leo Postman in which the colour was reversed on some of the cards, people saw, for example, a black 4 of hearts as either a 4 of spades or a normal 4 of red hearts[1]. Their expectation about what the cards 'should' look like led them to erroneously report the actual reality. I often wonder how many people (often highly trained and competent astronomers, geologists and map makers) have looked at the landscape of Mars yet they have not seen or suspected anything unusual about the placement of the major mountains. Perhaps some have, but were not allowed to report it (e.g., NASA employees), or were too afraid of ridicule to say anything.

When I first looked at the splendid map on the inside cover of *Visions of Mars* by Olivier de Goursac[2], I was struck by the patterns that seemed to leap out from the page in the layout of the mountains of Mars, particularly Olympus Mons and the Tharsis Montes (Ascraeus Mons, Pavonis Mons and Arsia Mons). These seemed to form a bisected equilateral triangle, i.e., a triangle with 3 equal sides divided into 2 equal parts by an imaginary line that could be drawn between Olympus Mons and Pavonis Mons (see Fig. 1.1). Then, looking at Elysium Mons and its 2 close companions, Hecates Tholus and Albor Tholus, I noted that the latter 2 peaks seemed to lie on the same meridian and therefore could be used as a pointer to the North Pole (Fig. 1.2). An outrageous thought crept into my mind which I found impossible to shake. Could it be that these mountains were not naturally formed, but actually laid out by a highly sophisticated race of beings who were in possession of an unimaginably powerful technology capable of creating the most massive mountains in the solar system? I, like others, have only thought in terms of mountains being formed due to natural forces at work, never due to engineering by intelligent beings.

One of the biggest reasons for the inability of others to see the mountain patterns and other major signs of artificiality has been the lack of accurate coordinates. NASA coordinates are very misleading since they use the wrong prime meridian and they are not correctly located according to the design of the original architects. It was only through the discovery of survey markers which still remain visible on the Martian surface that I was able to calculate out the correct coordinates with accuracy sufficient enough to measure out the proper distances and angles of the patterns that reveal the artificiality of the Martian topography. I was greatly aided in this by the availability of satellite maps of the Martian terrain supplied by the USGS Astrogeology Research Program. The modern personal computer and software have also made it possible to do calculations and graphics that would otherwise have been beyond the reach of a single investigator working outside of a research environment.

What I eventually found went far beyond my initial observations, and has forced me to start rethinking everything our culture has taught regarding our own origins, our solar system, and the universe itself. I now invite the reader to follow my journey of discovery. But be prepared to leave forever the 'security' of the total world view to which you now subscribe.

References

1. *On the Perception of Incongruity: A Paradigm. J. S. Bruner and L. Postman. The Journal of Personality, vol. 18, 1949, pp. 206-223.*

2. *Visions of Mars. Olivier de Goursac. Translated from the French by Lenora Ammon. Harry N. Abrams, Inc. New York. 2005.*

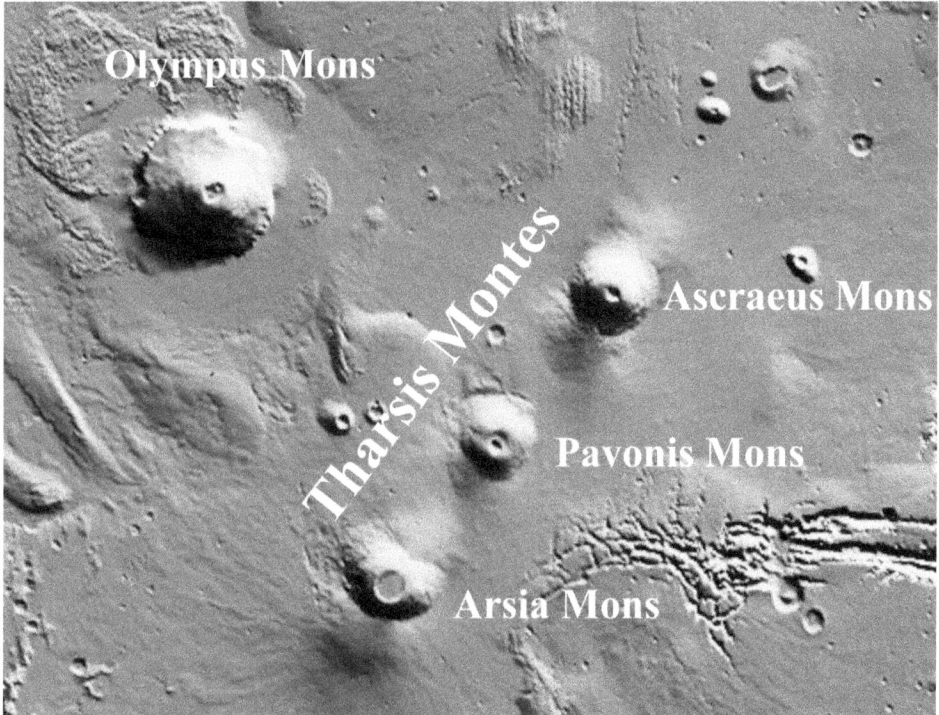

Fig. 1.1: *Olympus Mons and the Tharsis Montes. This mountain group gives the impression of forming an equilateral triangle split in 2 by an imaginary line between Olympus Mons and Pavonis Mons. MOLA Science Team. Courtesy NASA/JPL-Caltech.*

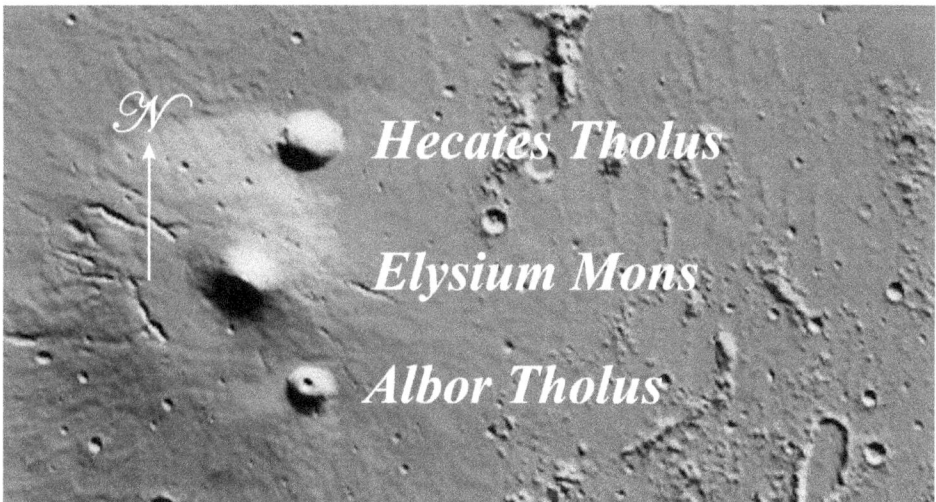

Fig. 1.2: *Elysium group of mountains. Hecates Tholus and Albor Tholus appear to lie on the same meridian. MOLA Science Team. Courtesy NASA/JPL-Caltech.*

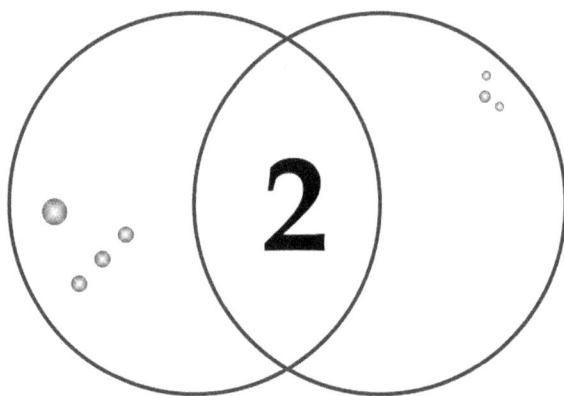

An Orientation Tour of Mars

For many of us who are not familiar with Mars other than what we have seen and heard of in the mainstream media, probably what first comes to mind is the vision of a lifeless, rocky and cratered planet with reddish coloured 'soil'. Not much attention has been given to the great diversity of geological features that this world possesses and hence, a quick tour of Martian geography is probably in order before I go on to discuss the main focus of this book which is the mountains of Mars. To assist this, I have labelled a global map of Mars which appeared on the cover of Science magazine in May of 1999 (see Fig. 2.1). The map is coloured coded according to elevation above datum (a measure of 0 elevation which is defined as the elevation at which the Martian atmospheric pressure is 6.1 millibars). Conveniently, the colour chosen to represent an elevation of -4 to -5 km is blue which gives the impression of water in regions where water was most likely to have occurred in ancient times. Green areas representing an elevation of about -2 km would also be expected to have been covered with water during the wettest epochs.

The most obvious feature which this map vividly demonstrates to us is the image of a vast "ocean" in the northern hemisphere that occupies the upper third of the planet and surrounds and even covers the North Pole. Coming off this "ocean" are 3 very large bays which extend southward close to the equator and are approximately equally spaced around the globe. Now empty the "ocean" of all its imaginary water, and you have

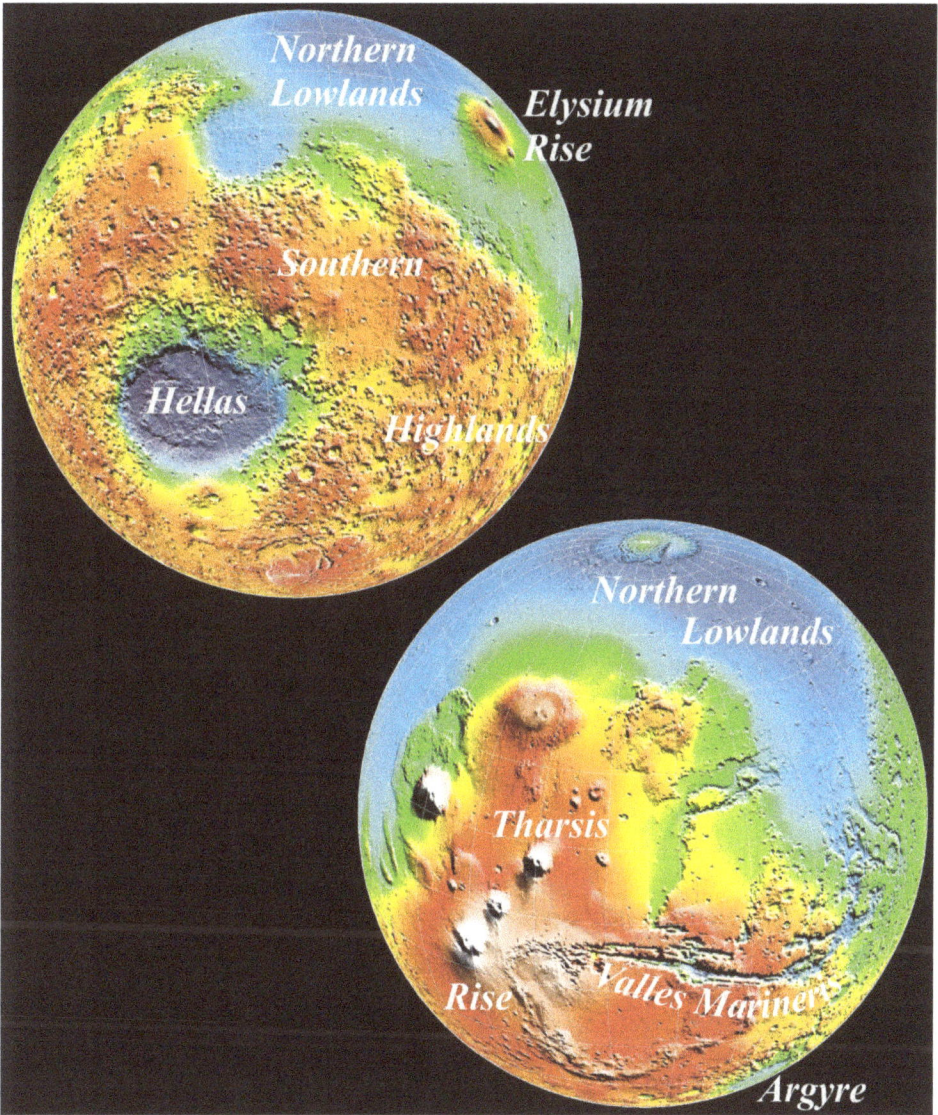

Fig. 2.1: *Mars Orbiter Laser Altimeter (MOLA) relief map showing the principle geographical features of Mars. Colour is based on elevation. With reference to datum (elevation where atmospheric pressure = 6.1 millibars), approximately: purple is <-5 km, blue is -4 to -5 km, green is -2 km, yellow 0 km, orange 2 to 4 km, red 5 to 7 km, and white > 7 km. NASA/JPL/GSFC Courtesy NASA/JPL-Caltech.*

what exists today, a vast hollow in the northern hemisphere which is quite flat and smooth, with relatively few craters to mark it up. Although the "shoreline" of this "ocean" varies in elevation by as much as 2.5 km, the deviations in elevation may have been caused by geological upheavals in later times[1,2]. The entire depression lies about 4 kilometers below datum, and the low crater count in this region suggests that if it were the bed of a real ocean it would have existed as late as about 2 billion years ago[2]. Further evidence comes from recent data obtained with the gamma ray spectrometer instrument carried aboard NASA's Mars Odyssey orbiter which indicated the presence of ancient northern shorelines marked by areas of enriched potassium, thorium and iron[3]. Two shorelines were detected, a younger one about 10 times the size of the Mediterranean and an older one about twice that size.

With the southern hemisphere, the opposite picture is found. Most of it lies 1 - 3 kilometers above datum, and it is heavily cratered, suggesting that it represents very old terrain much like the surface of the moon. The other prominent feature of the southern hemisphere is the presence of 2 huge impact basins, Hellas and Argyre, lying between 30 and 60 degrees of southern latitude. These are giant deep depressions in the Martian terrain thought to have been caused by large asteroids striking the surface of Mars about 4 billion years ago. Argyre has a diameter of 800 km and descends to a depth of about 5 kilometers. Hellas, with a diameter of about 2000 km and a depth approximately 9 kilometers below the surrounding terrain[4], spreads over an area more than 6 times that of Argyre. The South Pole is normally covered with a cap of condensed carbon dioxide which increases in size during the winter and recedes during the summer. It has recently been discovered to contain vast quantities of water ice as well as frozen carbon dioxide[5]. At the North Pole, a surface layer of carbon dioxide condenses out of the atmosphere during the winter and evaporates during the summer leaving a lower layer of frozen water which remains all year round.

In the middle of the planet, just south and roughly parallel to the equator lies a vast system of canyons so huge and long that they totally dwarf our own Grand Canyon. The Valles Marineris system of canyons runs eastward for about 5000 km and goes as deep as 11 km below the surrounding plains[6]. The system then turns northward and connects to the northern lowlands. By comparison, the Grand Canyon only runs for 446 km and is about 1.6 km in depth. Waterfalls up to 4 km high have been photographed in the Echus Chasma which lies just north of Valles Marineris and runs northward[7]. They reveal the existence of surface liquid water in enormous quantities billions of years ago.

The final features on our brief tour of Mars are its major mountains and the bulges that they sit on. These mountains occur in 2 main groups – those associated with the Tharsis Rise and those lying on the Elysium Rise. The Tharsis Rise is an immense tract of terrain elevated to a peak of about 10 km and extending more than 8000 km across[8]. It covers an area of about 30 million square kilometers or about 3 times the size of the

United States, and occupies about 20% of the planet's surface. The Tharsis Rise hosts 3 of the largest mountains on Mars as well as the Valles Marineris system.

The most obvious Martian mountain is Olympus Mons (Fig. 2.2), sitting just off the western edge of the Tharsis Rise. Everything about this volcanic mountain is gigantic. It has a median diameter of about 625 km and rises about 25 km above the surrounding plane or 22 km above datum[9,10] (i.e., its base lies below datum). This is about three times the height of Mount Everest at 8.85 km. The overall shape of Olympus Mons is very different from what most people would normally expect a mountain to have. To give the reader an idea of just how it looks, a drawing of the approximate mean cross section of this mountain based on data from McGovern and Morgan[11] can be seen at the beginning of each chapter between the chapter number and title. It has the appearance more of a Frisbee than the conical shape of volcanic mountains with which we are more familiar. Geologists refer to it as a shield volcano which is a mountain that is formed from low viscosity lava (i.e., lava which flows readily). Actually there are plenty of shield volcanoes on earth such as the Hawaiian mountains of Mauna Lao and Mauna Kea, but they are small in comparison to this Martian giant. Incidentally, my drawing of the cross section of Olympus Mons is about 1/5 of one millionth of actual size.

Just to the east of Olympus Mons are 3 more huge mountains (termed the Tharsis Montes) arranged approximately on a straight line running from southwest to northeast on the Tharsis Rise. From north to south, with heights above the surrounding region[12] (first number) and elevations above datum[6] (second number) given in brackets, these are Ascraeus Mons (13.9, 18.2 km), Pavonis Mons (8.2, 14.1 km) and Arsia Mons (9.3, 17.8 km).

The second major group of Martian mountains is located on the Elysium Rise, about a quarter of the planet to the west of the mountain group on the Tharsis Rise. The Elysium Rise is about 1700 by 2400 km in size and has an elevation approximately 5 km above datum[13,14]. Elysium Mons is the largest of this group of 3 with a height of 12.5 km above the surrounding plain and an elevation of about 16 km above datum[15]. Its diameter is about 240 km. North of Elysium Mons is Hecates Tholus which has a 180 km diameter and a height of 5.3 km[16]. South of Elysium Mons is Albor Tholus with a height of about 4.5 km and diameter of 160 km[17].

Water has now been discovered to still exist in large quantities on Mars. It has been observed on 2 occasions to have risen to the surface in the liquid state, each time forming a gully as it ran down the inside surface of a crater[18]. In Jan-Feb of 2004, the European Space Agency Mars

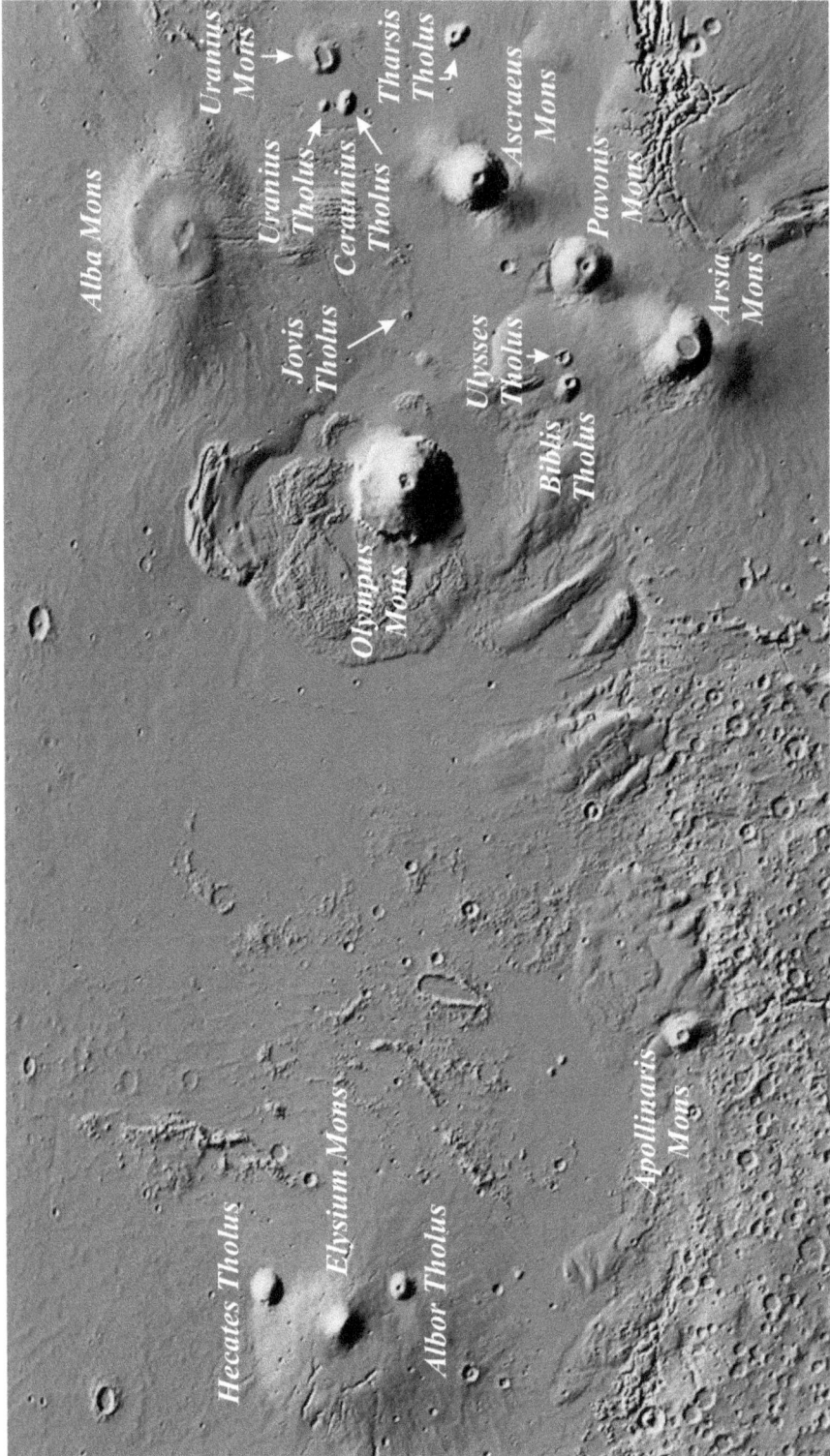

Fig. 2.2: *The mountains of Mars discussed in this book. Pavonis Mons lies just north of the equator and Hecates Tholus just north of 30 degrees latitude. This image covers almost 40% of Martian longitude. MOLA Science Team Courtesy NASA/JPL-Caltech.*

Express satellite detected hundreds of square kilometers of permafrost (frozen water mixed with soil) around the South Pole[5]. Using the Mars Advanced Radar for Subsurface and Ionospheric Sounding (MARSIS) instrument the Mars Express satellite was able to measure the quantity of frozen water buried beneath the South Pole to be equivalent to a liquid layer covering the entire planet to a depth of about 11 meters[5]. More recently the Phoenix Mars Lander found water ice beneath the soil under the vehicle[19]. It even detected snow falling from the Martian clouds[20].

So Mars is actually a very extraordinary planet by several measures. It has the remnants of a giant northern ocean which is postulated to have been present during ancient times. It possesses an impact basin (Hellas) that could hold enough water to cover the entire planet to a depth of about 195 meters, a canyon system that makes our own Grand Canyon look like a creek, and Olympus Mons, the tallest mountain in the solar system. And if this is not enough, consider that the Tharsis Rise is itself like a giant gentle-sloped mountain with an elevation equal to Mount Everest and an expanse which would cover more than Russia and Canada combined. No wonder this planet has attracted so much of our interest and space exploration funding.

References

1. *Evidence for an ancient Martian ocean in the topography of deformed shorelines. J. Taylor Perron, Jerry X. Mitrovica, Michael Manga, Isamu Matsuyama & Mark A. Richards. Nature 447, 840-843 (14 June 2007).*

2. *Strong evidence that Mars once had an ocean. Robert Sanders, Media Relations 13 June 2007. UC Berkeley News. http://berkeley.edu/news/media /releases/2007/06/13_mars.shtml*

3. *More evidence points to past oceans on Mars. November 2008. Ivan Semeniuk reporting work by James M. Dohm, Victor Baker, William Boynton et al. http://www.newscientist.com/article/dn16063-more-evidence-points-to-past-oceans-on-mars.html*

4. *Mars Express HRSC - study of a paleolake chain in Northern Promethei Terra, Mars.   H. Lahtela, V.-P. Kostama, J. Korteniemi, J. Raitala, G. Neukum and the HRSC Co-Investigator Team. Geophysical Research Abstracts, Vol. 7, 00602, 2005. http://www.cosis.net/abstracts/EGU05 /00602/EGU05-J-00602.pdf*

5. *Mars Express. Radar Gauges Water Quantity Around Mars' South Pole. 15 March 2007. http://www.esa.int/Our_Activities/Space_science/Mars_Express /Mars_Express_radar_gauges_water_quantity_around_Mars_south_pole*

6. *Visions of Mars. Olivier de Goursac. Translated from the French by Lenora Ammon. Harry N. Abrams, Inc. New York. 2005.*

7. *Mars Express. Glacial, Volcanic and Fluvial Activity on Mars: Latest Images. Feb 25, 2005. http://www.esa.int/Our_Activities/Space_science /Mars_Express/Glacial_volcanic_and_fluvial_activity_on_Mars_latest_images*

8. *Gravity Studies of the Tharsis Area on Mars. Peter Janle and Ercan Erkul. Earth, Moon and Planets, 53: 217-232, 1991.*

9. *How Volcanoes Work. Volcanism on Mars. http://www.geology.sdsu.edu /how_volcanoes_work/mars.html*

10. *Flying Above Mount Olympus. Astrobiology magazine, Feb. 15, 2004. http://www.astrobio.net/exclusive/832/flying-above-mount-olympus.*

11. *Spreading of the Olympus Mons Volcanic Edifice, Mars. P.J. McGovern and J.K. Morgan. Lunar and Planetary Science XXXVI (2005). http://www.lpi.usra.edu/meetings/lpsc2005/pdf/2258.pdf*

12. *Heights of Martian Volcanoes and the Geometry of their Calderas from MOLA Data. P. J. Mouginis-Mark and K. J. Kallianpur. Lunar and Planetary Science XXXIII (2002).   http://www.lpi.usra.edu/meetings /lpsc2002/pdf/1409.pdf*

13. *Martian Volcanoes. Calvin J. Hamilton. http://www.solarviews.com/eng /marsvolc.htm*

14. *Initial Results from MGS Pass 3: MOLA Capture Orbit Calibration Data. Press Release Figures from First Mars Global Surveyor Science Press Conference, October 2, 1997. http://tharsis.gsfc.nasa.gov/pass3.html*

15. *Photojournal, Jet Propulsion Laboratory, California Institute of Technology. PIA01457: Elysium Mons Volcanic Region. http://photojournal.jpl.nasa.gov /catalog/PIA01457*

16. *ESA Mars Express March 1, 2004. Hecates Tholus Volcano in 3D. http://www.esa.int/Our_Activities/Space_Science/Mars_Express /Hecates_Tholus_volcano_in_3D*

17. *ESA Mars Express. Dust Falling into Caldera of Albor Tholus. http://sci.esa.int/mars-express/34520-dust-fall/*

18. *NASA Images Suggest Water Still Flows in Brief Spurts on Mars. NASA website. December 6, 2006. https://www.nasa.gov/mission_pages/mars/news/mgs-20061206.html*

19. *NASA Phoenix Mars lander Confirms Frozen Water. June 6, 2008. http://www.nasa.gov/mission_pages/phoenix/news/phoenix-20080620.html*

20. *NASA Mars Lander Sees Falling Snow, Soil Data Suggest Liquid Past. September 29, 2008. http://www.nasa.gov/mission_pages/phoenix/news/phoenix-20080929.html*

3

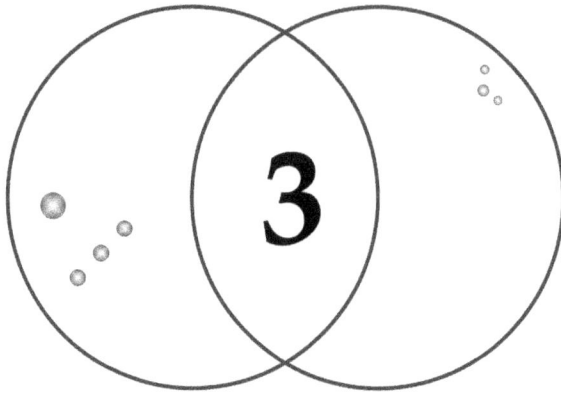

Working Out Technique

Measuring the distance between mountains for the purpose of evaluating the likelihood of artificial patterns is not as easy as it would first seem. First of all, the huge lengths involved require that the spherical nature of the Martian planet be taken into account. Secondly, the occurrence of several eruptions over the ages has in many instances covered over the original caldera (crater formed at the top of a mountain after a volcanic eruption or collapse of the cone) and may have shifted the position of the peak by several kilometers. One must also decide on what coordinate system to use, and how to translate positions on a map into accurate coordinate numbers.

Spherical Geometry

Spherical geometry offers 2 main choices for measuring the distance and bearing (clockwise or counterclockwise angle from due north - see Fig. 3.1) of a line joining any 2 points on a spherical surface: either great circles or rhumb lines. A great circle is the intersection of the surface of a sphere with a plane passing through its centre (Fig. 3.2). Any great circle divides the surface of the sphere into 2 equal halves. Examples of great circles are the earth's equator and meridian lines (shortest distance lines running along the surface of the sphere which connect the North Pole with the South Pole and mark out the longitude coordinates of locations on a

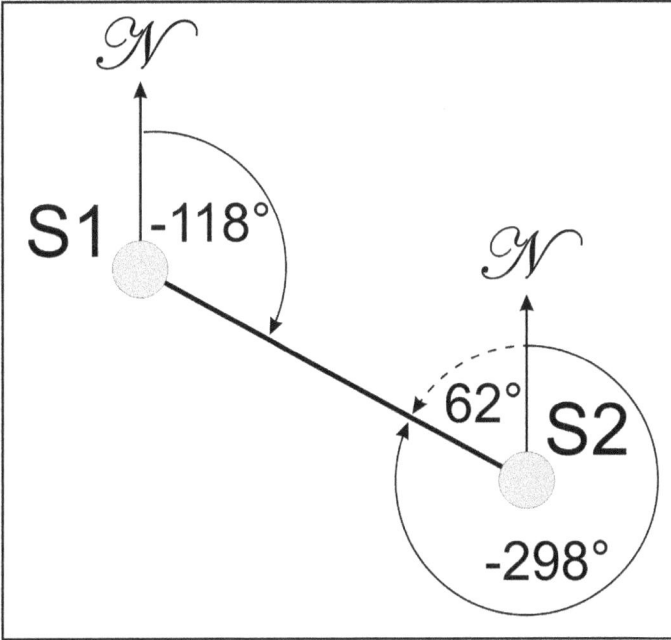

Fig. 3.1: *The bearing of a path joining 2 sites (S1 and S2) on a planetary surface is the angle formed with a line pointing due north. I will use negative angles in this book to denote clockwise bearings and positive angles to denote counterclockwise bearings. Note that the clockwise bearing angle differs by 180° when measured along the path in the direction S1 to S2 compared to the direction S2 to S1.*

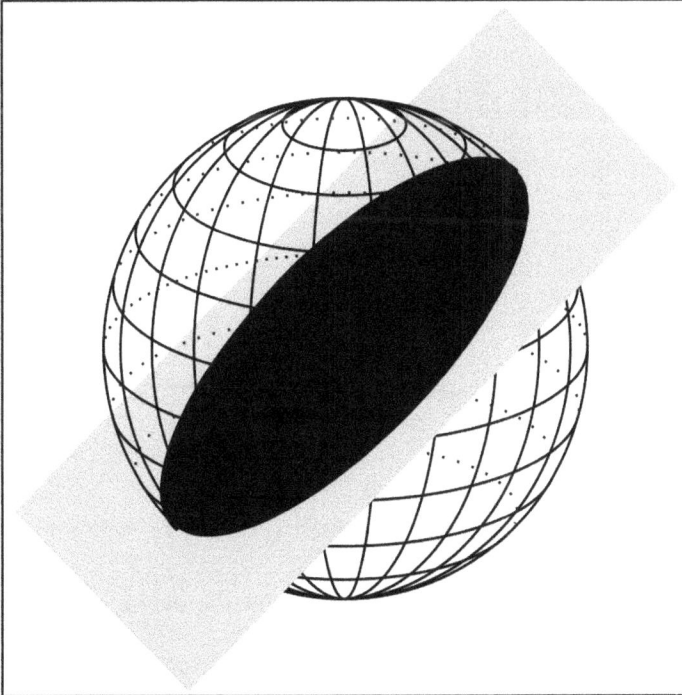

Fig. 3.2: *A great circle is the intersection between the surface of the sphere and a plane passing through the centre of the sphere. In this diagram, the outside perimeter of the black ellipse is a great circle. All meridian lines (shortest distance lines running between the poles) also mark out great circles.*

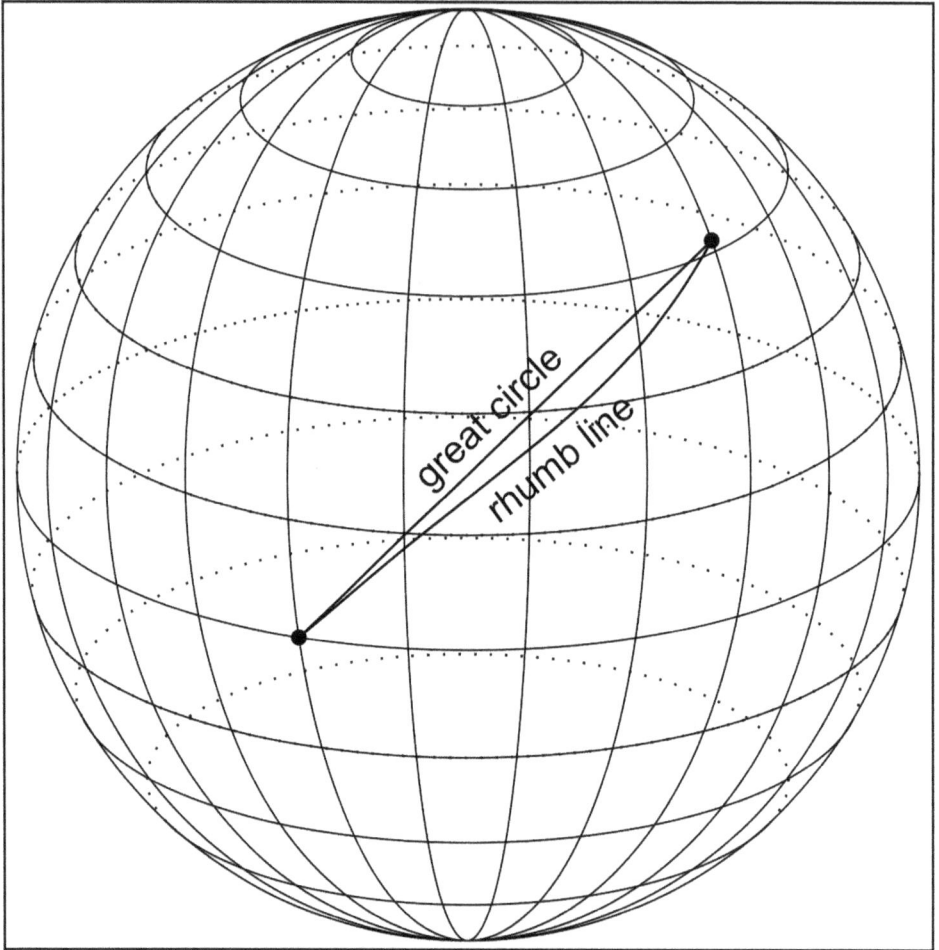

Fig. 3.3: *Comparison of a great circle path between 2 sites with a rhumb line path on a planetary sphere. The great circle path is the shortest distance between the 2 locations and appears as a straight line when viewed directly overhead from outer space even though it follows the curvature of the planetary surface. Its bearing is continually changing along the path as can be seen from its intersections with the meridian lines. The rhumb path is longer and curved, but maintains its bearing over the entire distance.*

planet). A great circle that passes through any 2 different points on a planet provides the shortest possible path between them. However, the bearing of this path varies over its length unless the 2 points lie on the equator or on the same meridian line. A rhumb line on the other hand is the line formed by a path maintaining a constant bearing between the 2 points (Fig. 3.3). Rhumb lines were often used by navigators to simplify the process of maintaining course on ocean journeys despite the expense of taking a longer path. They also needed to be used prior to the invention of clocks accurate enough to enable longitude to be calculated at sea.

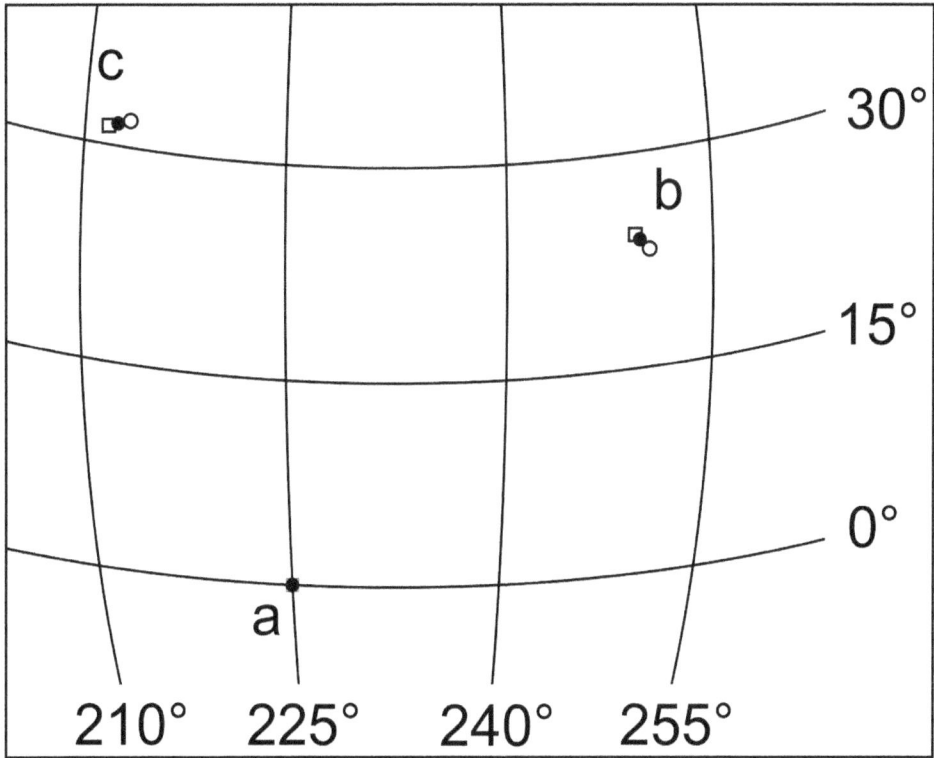

Fig. 3.4: *A hypothetical equilateral triangle created by virtual sites placed on the surface of Mars within the latitude boundaries of the Martian mountain sites under study. The length of each side of the triangle is set to 2000 km, and the bearing of the side between sites a and b is intended to be -45°. The filled circle symbols represent sites placed using rhumb lines. Open circle symbols represent the sites placed with great circles, with the initial bearing (at site a) of the great circle between sites a and b set to -45°. The open square symbols represent the sites also placed with great circles, but now the great circle between sites a and b is set so that the mean of its bearings at sites a and b equals -45°. Site a is located on the equator and is positioned at the same location for all 3 models. See text for discussion.*

If the mountains on Mars were deliberately laid out by hypothetical architects to create a geometric figure, they would have to decide on how best to represent the figure on a spherical surface. Should they measure out the distances between sites according to great circles, or according to rhumb lines? If distance were the only consideration, then great circles would give the best result. However, if you wanted to encode bearing information in your geometric figures, using rhumb lines would be the only way to go. This is illustrated in Fig. 3.4 in which I constructed 3 hypothetical equilateral triangles on Mars with each side equal to 2000 km. One triangle was constructed using rhumb lines, and the other two using great circles[1]. The triangles were placed within the latitude boundaries of all the mountains which I intended to study since I wished to simulate the actual conditions as closely as possible. I wanted the side between sites a and b to have a bearing of -45°. This was easily accomplished with the first triangle by using rhumb lines to lay out the site positions. When I wanted to construct a triangle using great circles, however, I had a dilemma. Should I set the initial bearing of this side, or the mean of its initial and final bearings to -45° ? The first solution caused the triangle to be rotated clockwise from the rhumb triangle, with the eastern site positioned about 58 km away from that of the rhumb triangle. The final bearing of this side was -50.25° instead of -45°. The second solution caused the triangle to be rotated counterclockwise with the eastern site positioned about 29 km from that of the rhumb triangle. The initial bearing of its side was -42.36° and the final bearing was -47.64°. This demonstrates that great circle solutions give poor representations of angles. If the hypothetical architects wanted to encode the size of an angle in a geometric figure, it would only make sense to an observer if the angle maintained its value over distance. For this reason, I started out creating models and measuring site-to-site distances using rhumb lines instead of great circle lines. This decision turned out to be the correct one for the geometric figure created by Olympus Mons and the Tharsis Montes which is studied in Chapter 5.

Calculating Coordinates

There are now 2 coordinate systems used in the mapping of Mars. The planetographic system with longitude increasing to the west was used on maps produced before approximately 2002. The planetocentric system with longitude increasing to the east is used on maps produced after approximately 2002. Sometimes both systems appear on the same maps. The planetocentric system was initially selected for this study on the basis that it is now the current standard. This ultimately turned out to be the best choice for reasons which will be described later in Chapter 5.

I used a greyscale MOLA (Mars Orbiter Laser Altimeter) map (http://mars.jpl.nasa.gov/mgs/sci/mola/mar05-2000rel/ced2880x1440.jpg) for low resolution initial models. It did not have a grid superimposed on it, but since both longitude and latitude were linear, calculations of approximate coordinates were simple.

For making careful mountain and crater coordinate measurements, I used a higher resolution (1:5 million-scale) series of 30 coloured MOLA maps which divided the surface of Mars into 30 separate regions. These maps were colour coded according to elevation; I downloaded them from the Internet (Courtesy of the USGS Astrogeology Research Program, http://planetarynames.wr.usgs.gov/Page/mars1to5mMOLA)*. The MOLA maps were produced from the data obtained from the Mars Orbiter Laser Altimeter (hence, the abbreviation MOLA), an instrument aboard the Mars Global Surveyor (MGS) spacecraft which was launched on Nov. 7, 1996. It achieved research orbit on Sept 12, 1997, and stopped functioning at the end of June, 2001. During that time, the MOLA instrument made approximately 1 billion elevation measurements of the planet's surface. Each measurement was made by transmitting an infrared laser pulse to the planet surface and measuring the time it took to receive the reflected pulse by the MGS spacecraft.

On the coloured MOLA map dataset (the set of 30 maps dividing the surface of Mars into 30 separate regions) which I used for making careful mountain and crater coordinate measurements, both planetographic and planetocentric coordinate grids were present, and I used the latter grid for coordinate determinations as discussed above. On these maps, the latitude direction is nonlinear. Hence, a fifth order polynomial based on the positions of the latitude grid lines overlaying a particular map was used to calculate latitude values. The longitude dimension for maps of regions lying between latitudes of 30° S and 30° N was linear since they are Mercator projection maps. Calculating longitude values was therefore a simple procedure on these maps.

On the coloured MOLA maps covering regions between 30° N and 65° N, meridian lines overlaying the map were not parallel because the maps are Lambert conformal conic projection maps, and so required a more tedious procedure for longitude measurements. For these maps, I extended northward the meridian lines overlaying the map to a common point of intersection and used this point as the centre of map curvature. I then based longitude measurements on the number of degrees that an extended meridian line had to be rotated about this centre in order to precisely lie on the site of interest. Latitude measurements for these maps were based on a fit to a fifth order polynomial similar to the maps between latitudes of 30° S and 30° N.

The resolution of the coloured MOLA maps ranged from ±0.33 to ±0.47 km since the side length of one of the tiny square pixels composing the map is 0.667 km at the equator, giving a pixel diagonal size of 0.94 km. The absolute accuracy (the closeness to the true value) of the MOLA measurements upon which the map pixels are based is of the order of 100 meters horizontally and 1 meter vertically[2].

Drawing Geometric Figures on Maps

As mentioned above, I used a low resolution greyscale MOLA map for initial testing of geometric models for Olympus Mons and the giant mountains on the Tharsis Rise. Since it was an equidistant cylindrical map, rhumb lines were well approximated by straight lines within ±20 latitude degrees of the equator where these giant mountains lay.

The higher resolution (1:5 million-scale) series of 30 coloured MOLA maps used for more precise coordinate measurements were either Mercator or conical projection maps as discussed above. A very convenient feature of Mercator projection maps is that rhumb lines drawn on them are straight, and suited my purposes very well for pictorial representation of the various mountain models. A straight line drawn to represent a rhumb line on a conical projection map is a distortion, however, and is to be considered only as an approximation of reality for illustration purposes.

Determining Original Mountain Positions

The possibility of multiple eruptions hiding the original caldera and causing a shift in the position of the caldera/peak required a certain amount of detective work to try to arrive at plausible candidates for the original coordinates of any given mountain. Even the more recent calderas often showed several different eruption centres within their bounds. It was first decided to consider all of the following as candidates for original mountain coordinates:

- the centre of any recent caldera or of vestigial traces of an ancient caldera.
- the centre of the entire mountain.
- the centre of any arc of a circle in the region of the mountain perimeter which could possibly have formed the boundary of an earlier version of the mountain.

It eventually became apparent that the latter 2 options gave the closest approximation to original mountain centres which were uncovered by a technique which will be revealed in Chapter 5.

Testing Hypotheses

The main purpose was to determine the intent of my hypothetical architects, not to impose a structure that was not there. I started out with the primary hypothesis that many of the mountains of Mars were artificially positioned in geometric patterns. My self-assigned task was to reveal the code that was supposedly embedded in the locations of these enormous objects. This required making decisions as to which mountains to group to test for geometric patterns. Hence, it was very much like trying to solve a puzzle produced by a brilliant game creator. If the primary hypothesis or any sub-hypothesis was wrong, it would become painfully obvious in short order. This did not happen with the primary hypothesis but occurred several times with the sub-hypotheses (e.g., certain geometric models). However, it only inspired more effort since the patterns which I saw did not seem likely to be simply the result of natural processes. Through much trial and error, breakthroughs eventually did occur which led to deciphering more and more pieces of the puzzle, and enabled a cobbling together of a more complete picture.

* At the time of printing, I could no longer locate the 1:5 million-scale coloured MOLA maps on the Internet. They have recently been replaced with THEMIS maps. Hopefully the MOLA maps will once again be made available from the USGS Astrogeology Research Program.

References

1. *The programming code which I constructed for calculating rhumb and great circle distances and bearings was based on equations found on the Chris Veness website: Movable Type Scripts. Calculate distance, bearing and more between Latitude/Longitude points. http://www.movable-type.co.uk /scripts/latlong.html*

2. *Mars. Randolf L. Kirk. Grids and Datums, Photogrammetric Engineering and Remote Sensing. pp. 1111-1126, October 2005. http://www.asprs.org/a/resources/grids/10-2005-mars.pdf*

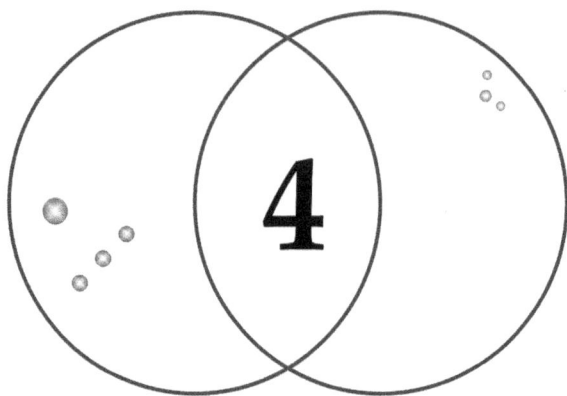

Olympus Mons and the Tharsis Giants

T he linearity of the position of the 3 Tharsis Montes has been well noted by many people. Current thinking attributes this to an assumed major fault line extending from the southwest to northeast which is now buried by volcanic outflows[1]. No explanation is given, however, for the fact that they appear to be equidistant from one another. Add to this the positioning of Olympus Mons and you cannot fail to notice that together, these 4 giant mountains have the look of a pattern to them. It is to these mountains then that I turned my attention to first, starting with simple models for testing and then discarding or improving them as more was learned.

Bisected Equilateral Triangle Model

My original hypothesis regarding the 4 large Martian mountains was that they seemed to mark out a bisected equilateral triangle, i.e., a triangle with 3 equal sides divided into 2 equal parts by a perpendicular line extending from Pavonis Mons to Olympus Mons. The easiest way to check this out in a preliminary fashion was to overlay such a triangle over a low resolution greyscale MOLA map (see Chapter 3) of that area of Mars. I then adjusted the size and positioning of the triangle to see if it could be aligned to the mountains, ignoring for the moment the distortions that the curvature of the planet would introduce over such long distances.

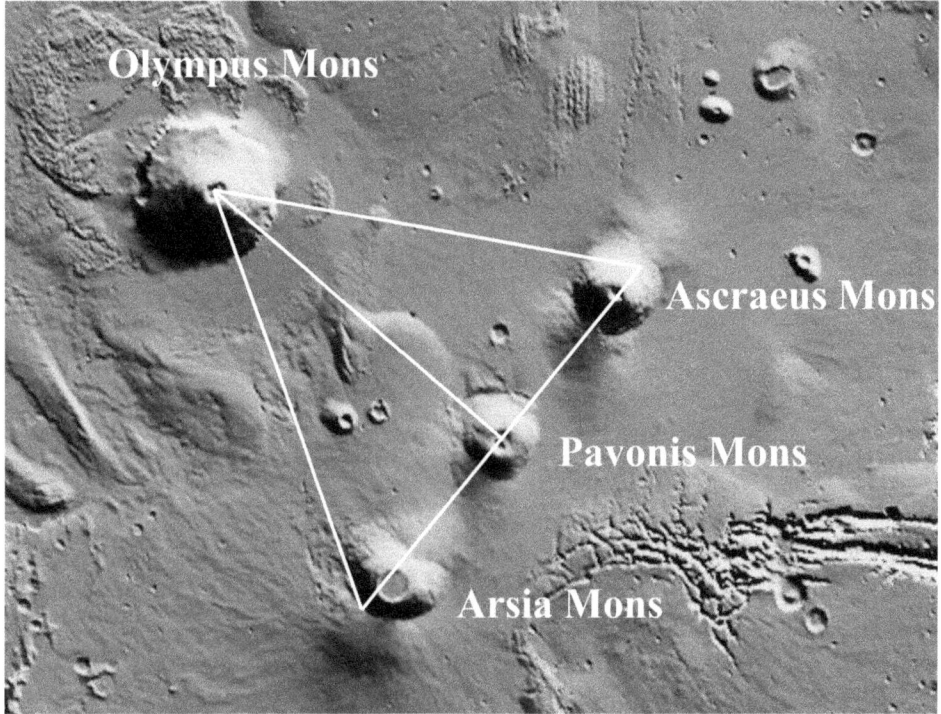

Fig. 4.1: *Equilateral triangle model. This model fails since the vertices at Ascraeus Mons and Arsia Mons do not fall in the central portion of the mountains. MOLA Science Team. Courtesy NASA/JPL-Caltech.*

Despite my best efforts to obtain a fit with an equilateral triangle, Fig. 4.1 shows that such a model was ultimately untenable. I was forced to make the side lying on the 3 Tharsis Montes stretch well beyond the centres of Ascraeus Mons and Arsia Mons in order to make the triangle large enough to reach the centre of Olympus Mons. Hence, the equilateral triangle hypothesis had to be abandoned very early on.

Bisected Isosceles Triangle Model

Although the equilateral triangle model failed, it did point the way to the next most obvious model which was a bisected isosceles triangle (a triangle with 2 equal sides). To test this out, I manipulated the dimensions of an overlaid isosceles triangle to see if a reasonable fit could be obtained. The results shown in Fig. 4.2 were reasonably successful this time. All the vertices of the triangle, including the base bisection point, could be made to simultaneously fall within the central regions of their respective mountains.

This was no trivial achievement. Here we had a symmetrical geometric

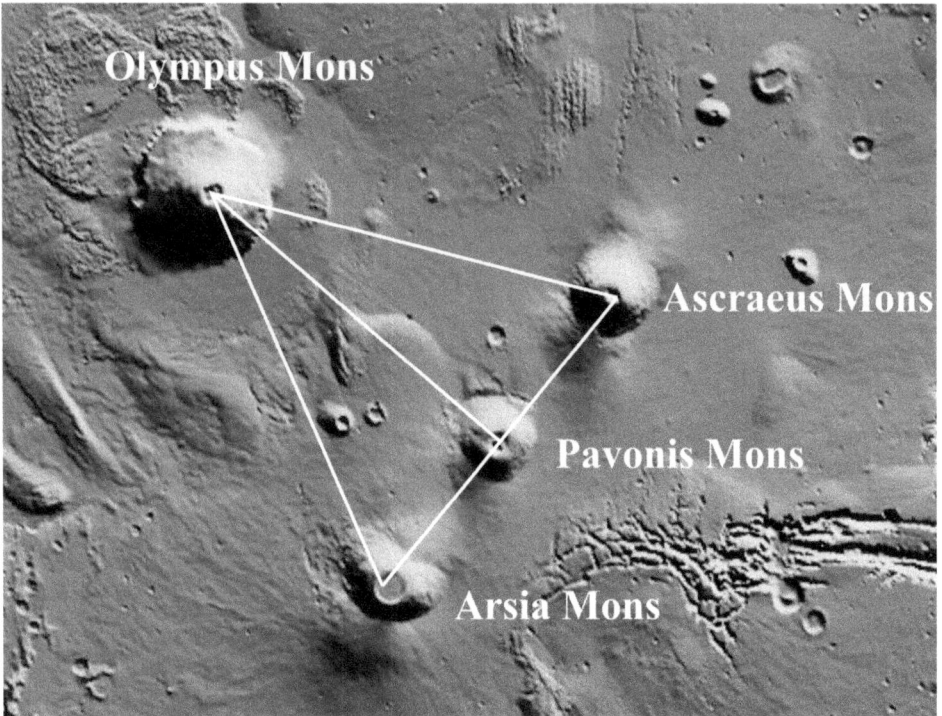

Fig. 4.2: *Bisected isosceles triangle model. This model succeeds in fitting all the vertices within the present calderas, including the bisection point of the base. Note, however, that at Ascraeus Mons and Arsia Mons the vertices are well off the centre of their respective calderas. MOLA Science Team. Courtesy NASA/JPL-Caltech.*

figure fitting itself to what was supposed to be the result of processes arising from natural causes alone. What was even more astonishing was that the length of the base turned out to be about 97 % of the height of the triangle, and hence, pointed to the possibility that a more accurate procedure would show them to be exactly equal. One other important thing that this exercise showed is that although all the vertices lay within the present calderas, the current caldera centres themselves were not going to fit the model exactly. This was only to be expected since it was obvious that the calderas had shifted many times over the course of time for each mountain as discussed in the previous chapter. If the original positions of the calderas of the mountains were laid out in a carefully planned pattern, these would now likely be hidden from view. The vertices of the isosceles triangle are particularly off-centre for the calderas of Ascraeus Mons and Arsia Mons. If an attempt is made to align the base of the triangle to the caldera centres of all 3 of the Tharsis Montes, the vertex at Olympus Mons ends up missing the caldera altogether. This is demonstrated in Fig. 4.3 which also shows that only the caldera centres of

Fig. 4.3: *Attempt to align the bisected isosceles triangle model to the 3 Tharsis Montes caldera centres. Only the caldera centres of Ascraeus Mons and Pavonis Mons can be properly aligned. MOLA Science Team. Courtesy NASA/JPL-Caltech.*

Ascraeus Mons and Pavonis Mons can be well fit when this is tried.

Bisected Isosceles Triangle Model Fitted to Mountain Centres

It was now obvious to me that if the mountains were artificially positioned to create a geometric pattern, the present caldera centres would not correspond to the original mountain centres. Therefore, an attempt was made to approximate the original coordinates of the 4 mountains by fitting a circle to the outer perimeter of each mountain. For Arsia Mons, since several perimeter options were available, I chose the most central one which surrounded the current caldera. The dimensions of a bisected isosceles triangle were then adjusted to fit the triangle to the circle centres instead of the caldera centres (Fig. 4.4). An excellent fit was achieved in this way, with the bisected isosceles triangle being only slightly misaligned to the circle centres of the 3 Tharsis Montes. Another striking result was that the triangle base for this model now turned out to be only

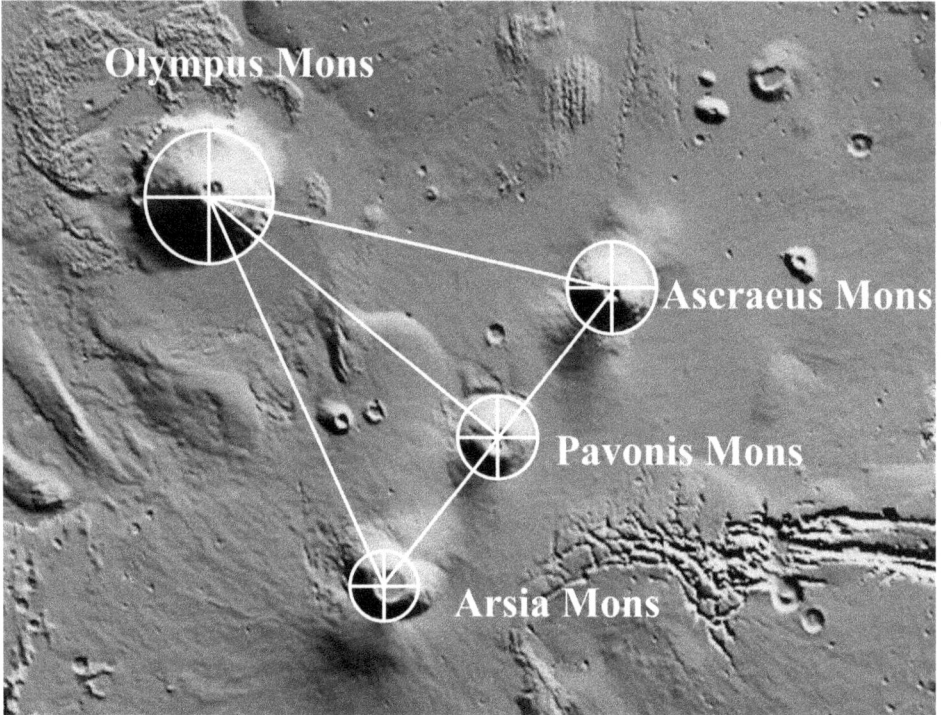

Fig. 4.4: *A bisected isosceles triangle model was sized and oriented to fit an approximation of the mountain centres. Circles were fit to the best guess of a mountain's outline. Since Arsia Mons seems to have wandered extensively during its history, an arc of a circle in the central portion of the southern edge of the mountain was fit. The intersection of 2 diameter lines in the shape of a cross for each mountain marks the centre of the circle surrounding the mountain. Note the closeness of fit to the mountain "centres" compared to the fit to the calderas in the model shown in Fig. 4.2. MOLA Science Team. Courtesy NASA/JPL-Caltech.*

about 0.8% shorter than the height of the triangle.

Sacred Geometry Model

The finding that the length of the base was virtually identical to the height in the previous model gave a very strong signal to look into the possibility that the mountains were deliberately positioned to form a pattern modelled on the principles of sacred geometry. But first I will give a brief description of sacred geometry for those who are not familiar with the subject.

The term 'geometry' etymologically means a measure of earth or land, such as a survey. However, it has been expanded to include a measure of all shapes and patterns in multiple dimensions, and even music and motion such as the movement of the planets. Geometry becomes sacred when it is used for spiritual or religious purposes. Whole numbers, ratios, multidimensional geometric shapes and patterns, harmonics and proportions, and special irrational number constants such as π (3.1416, the ratio of a circle's circumference to its diameter), e (2.7183, the base of the natural log) and φ (1.6180, the golden ratio) are expressed throughout the natural universe, and are all subject matter for sacred geometry. The irrational numbers run to an infinite number of decimal places and are therefore mystical quantities since we cannot completely determine their values.

Sacred geometry has been used by many religions and cultures to symbolically represent the Divine and all that has been created by the Divine. But it is more than that. It is also an acknowledgement that geometrical constructs are indeed a fundamental component of all of nature from particle physics to the macro universe. Geometry therefore reflects the deepest mysteries of reality. Sacred geometry has been developed to explain this in a systematic and reverential way. Hence, to create a piece of architecture or art based on sacred geometrical principles is to show reverence for all of creation and the ultimate Source behind it. To encode a measure of a planet or other heavenly body such as the sun or moon in a piece of work or symbol is to pay tribute to it either as a deity in itself or as the creation of a deity. Sacred geometry has also been used to summon or balance forces in nature such as in the use of pyramids and crystals for healing. Some have used it as a protection against evil. However, not everyone has used sacred geometry with good intent. Satanists have employed geometrical symbols in their rituals to try to summon up malicious entities or dark energies.

I now return to the most recent mountain model. By definition, all isosceles triangles have the special properties of 2 equal sides and 2 equal angles. The instance of the isosceles triangle which was fitted to an estimate of the mountain centres in Fig. 4.4 also had the special property of the height being virtually equal to the base. When we assign a value of 1 arbitrarily sized unit to half of the base and 2 of these units to the height (Fig. 4.5, left diagram), by using the Pythagorean Theorem it can be determined that the length of each of the 2 equal sides of the isosceles triangle will equal the square root of 5 (i.e., $\sqrt{[1^2 + 2^2]} = \sqrt{5}$). If we complete the rectangle with the dotted line in the left diagram of Fig. 4.5, a double square (2 squares of the same dimensions laid side-by-side) is constructed out of the left half of the isosceles triangle, and the side of our isosceles triangle becomes the diagonal of the double square. The double square is a geometric figure considered to be very sacred by past civilizations. We see it in Solomon's temple and in the King's Chamber of the Great

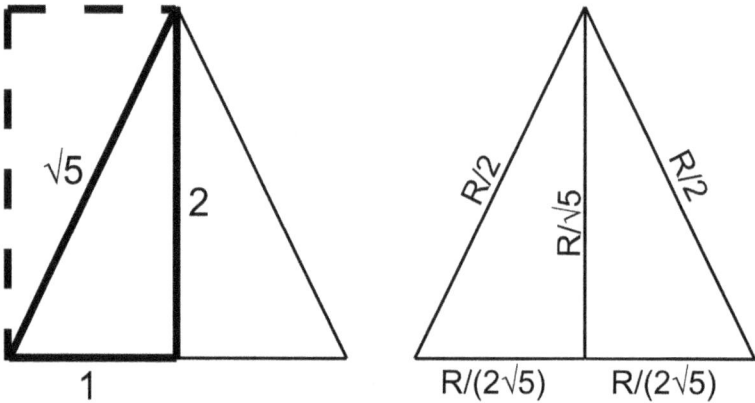

Fig. 4.5: *Left diagram: bisected isosceles triangle with the base equal to the height. If one-half of the isosceles base is given a value of 1 arbitrary unit, the height becomes 2, and the hypotenuse √5 by the Pythagorean Theorem. The dotted lines together with the right-angled triangle mark out a double square with the hypotenuse as the diagonal. For an identically shaped isosceles triangle on the right, by assigning one-half of the base a value of R/(2√5) km where R is the equatorial radius of Mars, by simple math the height becomes R/√5 km and the equal sides R/2 km. This becomes the sacred geometry model for fitting the layout of Olympus Mons and the 3 Tharsis Montes. See text for further explanation.*

Pyramid as examples. The reason for its importance is likely due to the fact that it contains the value of √5 in its diagonal, and √5 is an irrational number which encodes the golden ratio (φ = 1.6180). The value of √5 is linked to the golden ratio by virtue of the fact that it is equal to the sum of 1.6180 and 0.6180 (the inverse of the golden ratio). Also, (√5 + 1)/2 = 1.6180 and (√5 - 1)/2 = 0.6180.

Up to this point, I have made the measurements of the base and height of my model triangles in terms of centimeters only. This gave a somewhat distorted result since the model triangles did not take into account the nonlinearity of distance on the MOLA map resulting from the spherical nature of the planetary surface. I now proceeded to determine the actual coordinates for the vertices and base midpoint of the triangle fitted to the mountain centres from the model in Fig. 4.4. These were then used to calculate the rhumb line distances in kilometers for the sides, height and base of the fitted triangle (see Table 4.1). When this was done, I found that the height and base of the triangle differed by only 1 km or about 0.06% instead of the previously measured 0.8%. What is also very interesting is that the distances from the triangle vertex at Olympus Mons to those at Arsia Mons and Ascraeus Mons are only slightly greater than one-half the equatorial radius of Mars (R) which is 3396/2 = 1698 km. Note also that the distances from Olympus Mons to Arsia Mons and to Ascraeus Mons are not equal even though the sides of the model triangle

Table 4.1: *Distance measurements between Olympus Mons and the 3 Tharsis Montes, and between Ascraeus Mons and Arsia Mons. Mountain positions were taken to be the vertices and base mid-point of the triangle fitted to the mountain centre approximations of Fig. 4.4.*

Olympus Mons to Arsia Mons	1766 km
Olympus Mons to Ascraeus Mons	1714 km
Olympus Mons to Pavonis Mons	1569 km
Arsia Mons to Ascraeus Mons	1570 km

were made equal. Once again, this is because the surface of Mars is spherical, not flat. The average length of the 2 distances is only about 2.5% larger than one half the equatorial radius of Mars.

Dividing the equatorial radius of Mars by $\sqrt{5}$ gives us a value of 1519 km. This is only slightly less than the distance between the triangle vertex at Olympus Mons and the base midpoint at Pavonis Mons, and the vertex-to-vertex distance between Arsia Mons and Ascraeus Mons. The average length of these 2 distances is only about 3.3% larger than 1519 km. All of this suggests that decreasing the overall size of the isosceles triangle model by only 2.9% (the average of 2.5% and 3.3%) would result in an almost perfect fit to the base and height being $R/\sqrt{5}$ km and the 2 "equal" sides being a close approximation of $R/2$ km on a spherical surface. This would reduce the triangle in Fig. 4.4 by only about the width of the lines.

So now the latest model points to a second variable used in sacred geometry, namely the equatorial radius of the planet. To find that both R and $\sqrt{5}$ might be integrated into the layout of these mountains is very remarkable indeed. The next model to try was now very obvious. It would have the 2 equal sides of the isosceles triangle both set to the value $R/2$ km, and the base and height set to $R/\sqrt{5}$ km. Each half of the base would then be $R/(2\sqrt{5})$ km. These dimensions work out mathematically as well by conforming to the Pythagorean Theorem since $[R/(2\sqrt{5})]^2 + [R/\sqrt{5}]^2 = [R/2]^2$. This model is depicted in the diagram on the right in Fig. 4.5. However, before such a model could be applied, some trade-offs would have to be made in order to handle the unavoidable distortions caused by trying to fit a 2 dimensional model onto a spherical surface. This will be dealt with in the following chapter.

To summarize, what I found up to this point is that the locations of the central regions of the 4 giant mountains, Olympus Mons and the Tharsis Montes, can be fit with a relatively small amount of error by an isosceles

triangle model with the following properties:
- • 2 equal sides with lengths equal to R/2 km.
- • a base equal to the height, both having a value of R/$\sqrt{5}$ km.
- • a base that is bisected by Pavonis Mons.

What remained to be seen was whether or not the model was indeed the correct one. It could only be validated by having a very accurate measure of the original mountain centres. Was there some kind of marker of these centres that was still present after eons of volcanic eruptions? Or were the mountains just coincidentally placed by the forces of nature in a pattern that resembled my model but was in reality nothing more than an extraordinary curiosity? Although I found it hard to believe the latter, I knew that the deliberate placement of colossal mountains by an ancient intelligent civilization would be an extremely hard sell. At this point, I simply continued to make measurements of distances to other landmarks to see if more patterns would emerge. While doing this, a remarkable discovery was made which tipped the balance irrevocably towards the artificiality hypothesis. This then will be the subject matter of the following chapter.

References

1. *Cattermole, Peter John. Mars: the mystery unfolds. Oxford University Press, Inc., New York, N.Y., 2001.*

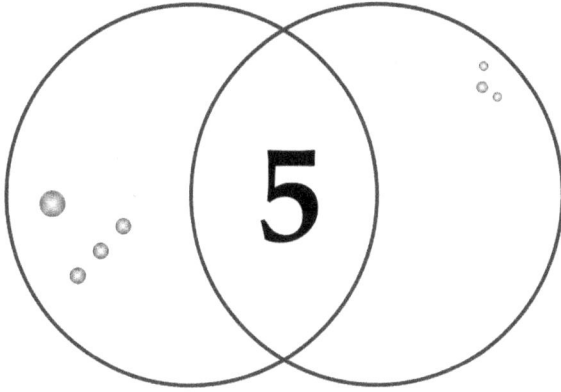

The Discovery of Survey Craters

T he modelling done in the previous chapter used a low resolution map of Mars. It was now appropriate to switch to the higher resolution 1:5 million-scale MOLA maps (0.667 km per pixel at the equator) downloaded from the Internet courtesy of the USGS Astrogeology Research Program (http://astrogeology.usgs.gov). The maps themselves are from the International Astronomical Union's Gazetteer of Planetary Nomenclature web page: http://planetarynames .wr.usgs.gov/Page/mars1to5mMOLA (see note at end of Chapter 3). These beautiful and accurate maps provide a dramatic view of Mars with the surface features colour coded according to elevation.

While exploring the distances between Ascraeus Mons and smaller mountains to the northwest such as Uranius Mons and Ceraunius Tholus, I decided to also measure the distances to the centres of some of the craters in between. Before I proceed further with this topic, however, I would like to discuss the concepts of accuracy, precision, resolution and the number of significant decimal places.

Accuracy is simply the closeness of a measurement to its true value. As stated in Chapter 3, the maps I used are based on MOLA measurements which have a horizontal accuracy of 100 meters and a vertical accuracy of 1 meter. Since each map pixel (the smallest discrete component of the map) likely involved several MOLA measurements each of which has error limits smaller than pixel size, map inaccuracy is highly unlikely to enter into the accuracy of my coordinate measurements in a significant way. Precision is a measure of the repeatability of a measurement. As a test of precision, I repeated the estimate of the coordinates of a large crater 10 times and determined that the standard deviation of my estimates was ±0.022 km or ±0.00037 degrees. This means that 68% of my estimates would lie within that range which, like MOLA measurements, is too small to have much of an impact on my measurements. Resolution is the fineness of detail that is present on the maps. As discussed in Chapter 3, resolution for these maps is somewhere between ±0.33 km (pixel width) and ±0.47 km (pixel diagonal) based on the dimensions of a pixel at the equator (0.667 km per side for these square shaped pixels). Thus, pixel size has a significant impact on measurement accuracy. Finally, the number of significant decimal places is the number of digits placed after the decimal point which are meaningful in a particular measurement. What is reasonable here? A coordinate measurement based on a single pixel would have a resolution of ±0.33 to ±0.47 km at the equator. Since one degree of longitude at the equator is equal to 59.27 km, the resolution of a single pixel would translate into ±0.0056 to ±0.0079 degrees. But as you will see, most of my coordinate measurements for sites on Mars were ultimately based on crater coordinate determinations, either directly or indirectly. A crater's coordinates were determined by fitting a circle to the crater perimeter and taking the coordinates of the centre of the circle to represent the position of the crater. Since the circle perimeter was fit to multiple pixels, the resolution of the measurement could now be considered to be improved by a factor of about 10, depending on the total number of pixels involved. This would then suggest that the resolution of a crater coordinate measurement can be as high as about ±0.0006 to ±0.0008 degrees. Hence the number of significant decimal places for a coordinate measurement could be considered to be 3 full places plus a 4th with partial significance. I therefore report all coordinate estimates to 4 decimal places. In terms of kilometers, the improvement in resolution by a factor of 10 would translate to about ±0.033 to ±0.047 km at the equator. It was therefore reasonable to report distance measurements to 2 decimal places.

The number of significant decimal places as discussed so far apply to the reporting of a central estimate by a single investigator. There is still the issue of personal bias to be considered. This is impossible to measure without the input from several trained investigators independently repeating the measurements according to their own biases. I therefore decided for practical purposes to treat all site location determinations as being somewhere within my best estimate ±0.50 km in any direction. Then, since a distance measurement depends on 2 site locations each with an assigned accuracy of ±0.50 km, I treated all distance measurements as having an accuracy of ±1.00 km. My best estimate of each distance was still reported to 2 decimal places to provide the centre point from which to apply the error limits. From the results that I achieved with the modelling of Olympus Mons and the Tharsis Montes which you will see in this chapter, this seemed to be a reasonable and practical procedure for handling distance accuracy.

Survey Craters for Ascraeus Mons

Getting back to my decision to measure crater distances from Ascraeus Mons, I was astonished to find that 2 of the craters (coordinates 260.0718° E 19.2776° N and 249.1186° E 17.9685° N) in a northerly direction from Ascraeus Mons were within a few kilometers of 540.52 km from where I determined the centre of this mountain to approximately lie (see previous chapter). This is a very significant distance because it incorporates not only the planetary radius (R) but also the extremely important irrational number π. The 540.52 km value is none other than the distance $R/(2\pi)$ km.

Was it possible that my hypothetical architects could not only position monstrous mountains, but could make enormous craters as well? I suppose blasting out craters would be a lot easier than engineering the location of an Olympus Mons. After all, our military weaponry is now capable of creating huge holes with nuclear explosives. But perfectly round holes 19 km or more in diameter? Hopefully for the sake of our own planet this is still a tall order.

What could be the purpose of creating craters at a distance from Ascraeus Mons that encoded R and π? Was it part of the mountain monument or could there be some other motivation? Then it struck me. Perhaps this was a method for locating the original peak of the mountain which shifted over time. There would only be a single location on Ascraeus Mons that would be exactly $R/(2\pi)$ km away from both craters, so these craters could be used as a survey tool. If my reasoning was correct, I now had a means to locate one of the vertices of my model triangle with a high degree of accuracy. All I had to do was to find the coordinates of the intersection point on Ascraeus Mons of 2 circles of radius $R/(2\pi)$ km, each centred on one of the craters. The site of interest was found to lie within the northern section of the present caldera, slightly west of centre (Figure 5.1) - a very credible candidate for the ancient mountain centre. The coordinates of this point were 255.5276° E 11.2791° N.

Naming these craters was a bit of a problem since no official name exists for them. I decided to give each of them the generic name of survey

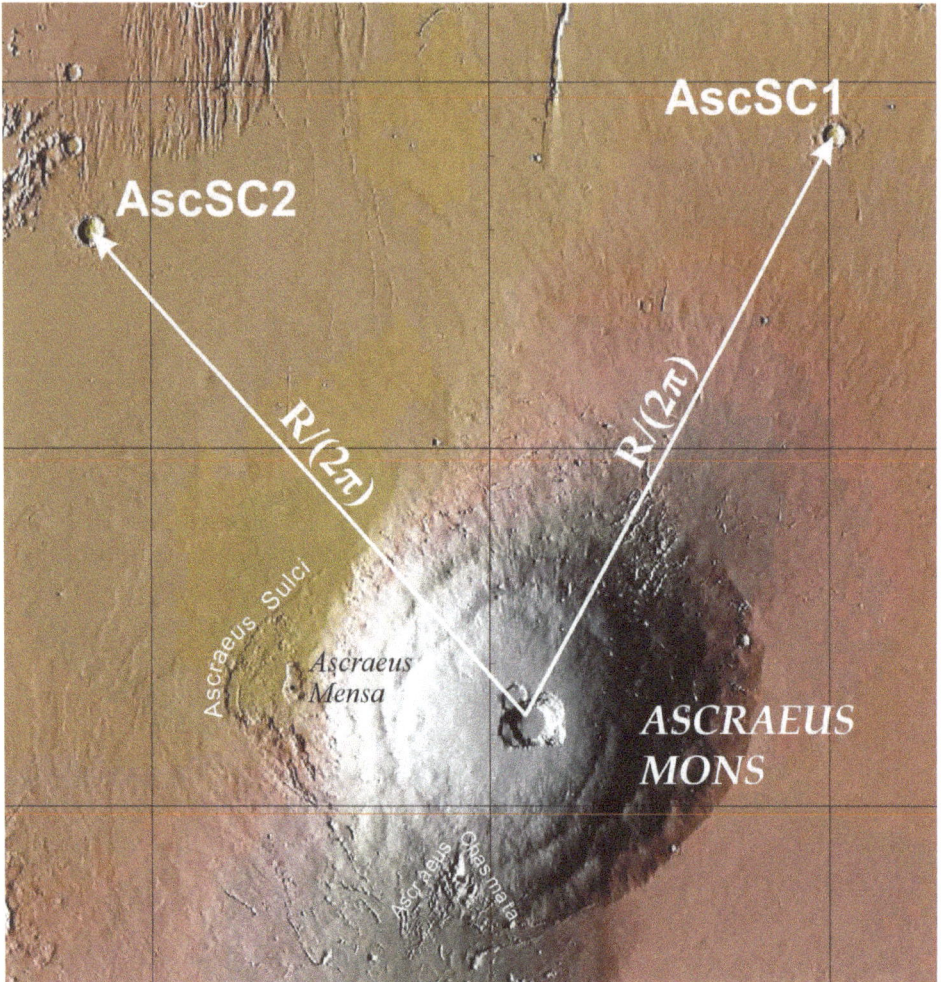

Fig. 5.1: *Survey crater model for locating the original centre of Ascraeus Mons. This centre was determined by locating the coordinates of the point which was R/(2π) km from each of the crater centres of AscSC1 and AscSC2. USGS Astrogeology.*

crater (SC). I further devised a system whereby they would be labelled with the first 3 letters of the mountain that they were associated with followed by the capital letters SC and a number. They were numbered in clockwise order starting from due north of the centre of the mountain. Hence, for Ascraeus Mons, I labelled the 2 craters AscSC1 and AscSC2.

At this stage in my analysis, a check of the other 6 craters in the area (there are not many craters near Ascraeus Mons of a reasonable size) did not reveal any encoding of sacred geometry in their distances from Ascraeus Mons. However, as I gained more experience with sacred numbers later on, I did find another crater with a more complex sacred distance which I will discuss in Chapter 9.

Survey Craters for Pavonis Mons

Was the survey crater find for Ascraeus Mons just a fluke? Could there be survey craters for the other mountains as well? I set about measuring crater distances from the mountain "centre" point for Pavonis Mons as determined in Chapter 4 from the centre of a circle fitted to surround the main mass of the mountain. These distances were compared to a set of likely sacred geometry candidates such as $R/(2e)$ or $R/(2\pi)$ km and checked for pairing.

Rather than a pair, no less than 5 craters were found whose distances from the Pavonis Mons mountain "centre" were suggestive of the same distance noted with the survey craters for Ascraeus Mons, namely $R/(2\pi)$ km. A sixth crater showed a distance close to the sacred distance of $R/(3\sqrt{5})$ km which is analogous to the distance of $R/(2\sqrt{5})$ km between the mountains of the base of the sacred geometry model triangle worked out in Chapter 4. I then named these craters PavSC1 to PavSC6 and determined the location on Pavonis Mons which gave the best fit to their sacred distances. Much later on, I discovered 2 more craters which had a distance of $R/(2\sqrt{5})$ km from this survey centre. As just mentioned, this is the same distance as that between the mountains of the base of the sacred geometry model triangle. I labelled the first of the pair as PavSC2a as it had a bearing angle between those for PavSC2 and PavSC3. The second had a bearing angle between those for PavSC5 and PavSC6 so I labelled it as PavSC5a. In this way I could avoid having to change the names that I previously assigned to the group of 6 survey craters.

I then recalculated the coordinates (247.1037° E 1.5993° N) of a point on Pavonis Mons which gave the least differences between the actual distances and the sacred geometry distances for all 8 craters. For fitting purposes, I used the squares of the differences rather than the actual differences since the squares are the most appropriate values to use in order to minimize the variation in a sample of data. The coordinates of the 8 survey craters are given in Table 5.1. Also listed are their approximate diameters. The deviations of the actual distance values from theoretical values were no more than 0.26 kilometers for all 8 craters from the newly surveyed location on Pavonis Mons.

Fig. 5.2 shows the location of the survey point on Pavonis Mons and the positions of the 8 craters used to determine the coordinates of the survey point. At this scale, some of the craters are not easily seen so I placed white circles around their perimeters to increase visibility. These survey craters varied from about 4 - 15 km in diameter and so many of them were considerably smaller than those found for Ascraeus Mons. This alerted to me the need to consider smaller craters as well as those of

Table 5.1: *Pavonis Mons survey crater coordinates and distances from the survey centre.*

Crater Name	Diam- eter (km)	Latitude	Longitude	Theoretical Distance (km)		Actual Distance (km)	Differ- ence (km)
PavSC1	8.7	2.5073 ° S	255.2465 ° E	$R/(2\pi)$ =	540.52	540.46	-0.06
PavSC2	4.7	3.3831 ° S	254.7409 ° E	$R/(2\pi)$ =	540.52	540.34	-0.18
PavSC2a	5.3	7.6929 ° S	255.9499 ° E	$R/(2\sqrt{5})$ =	759.41	759.57	0.16
PavSC3	14.7	2.6534 ° S	239.6955 ° E	$R/(3\sqrt{5})$ =	506.27	506.22	-0.05
PavSC4	4.0	7.4965 ° N	240.1250 ° E	$R/(2\pi)$ =	540.52	540.44	-0.08
PavSC5	4.6	10.5329 ° N	245.2474 ° E	$R/(2\pi)$ =	540.52	540.70	0.18
PavSC5a	5.2	14.2466 ° N	245.0620 ° E	$R/(2\sqrt{5})$ =	759.41	759.15	-0.26
PavSC6	10.5	10.6600 ° N	246.0514 ° E	$R/(2\pi)$ =	540.52	540.63	0.11

19 km diameter or more. A 4 km diameter crater would create a tight survey reference point without being invisible from quite a far distance.

Interestingly, the bearing of the line to PavSC3 is only 8 minutes less than 120 degrees from due north in the counterclockwise direction, and the line to PavSC4 is only 7 minutes more than e/π radians (= 49.576°) from due north in the counterclockwise direction[a]. The point on Pavonis Mons which was located with the aid of the postulated survey craters lay just beyond the northern edge of what appears to be the trace of a large diameter ancient caldera. It also was very close to the centre of an older outline of Pavonis Mons lying to the northwest of the present main bulk of the mountain. Hence it is a very credible candidate for the initial centre of Pavonis Mons. Also it should be noted that the latitude of this point (1.5993° N) is only about 1 minute of a degree less than the value for the golden ratio φ (1.6180).

Survey Craters for Arsia Mons

Searching for survey crater candidates for Arsia Mons proved to be a very challenging task. After measuring the distances from the mountain "centre" of no less than 42 craters located on the plains surrounding the mountain, 2 craters were identified that seemed to be at a "sacred" distance from Arsia Mons. Contrary to my previous experience with survey craters (except for 1 of the 6 original Pavonis Mons survey craters), however, this pair of craters represented 2 entirely different sacred distances, namely the now familiar distance of $R/(2\pi)$ km, and $R/(2\sqrt{5})$ km, the same distance as one-half of the base of the triangle of our postulated sacred geometry model. The coordinates of these 2 craters were 247.8850° E 8.0243° S and 241.6534° E 20.5837° S. Although different in sacred distance, they were similar in size, with the eastern crater

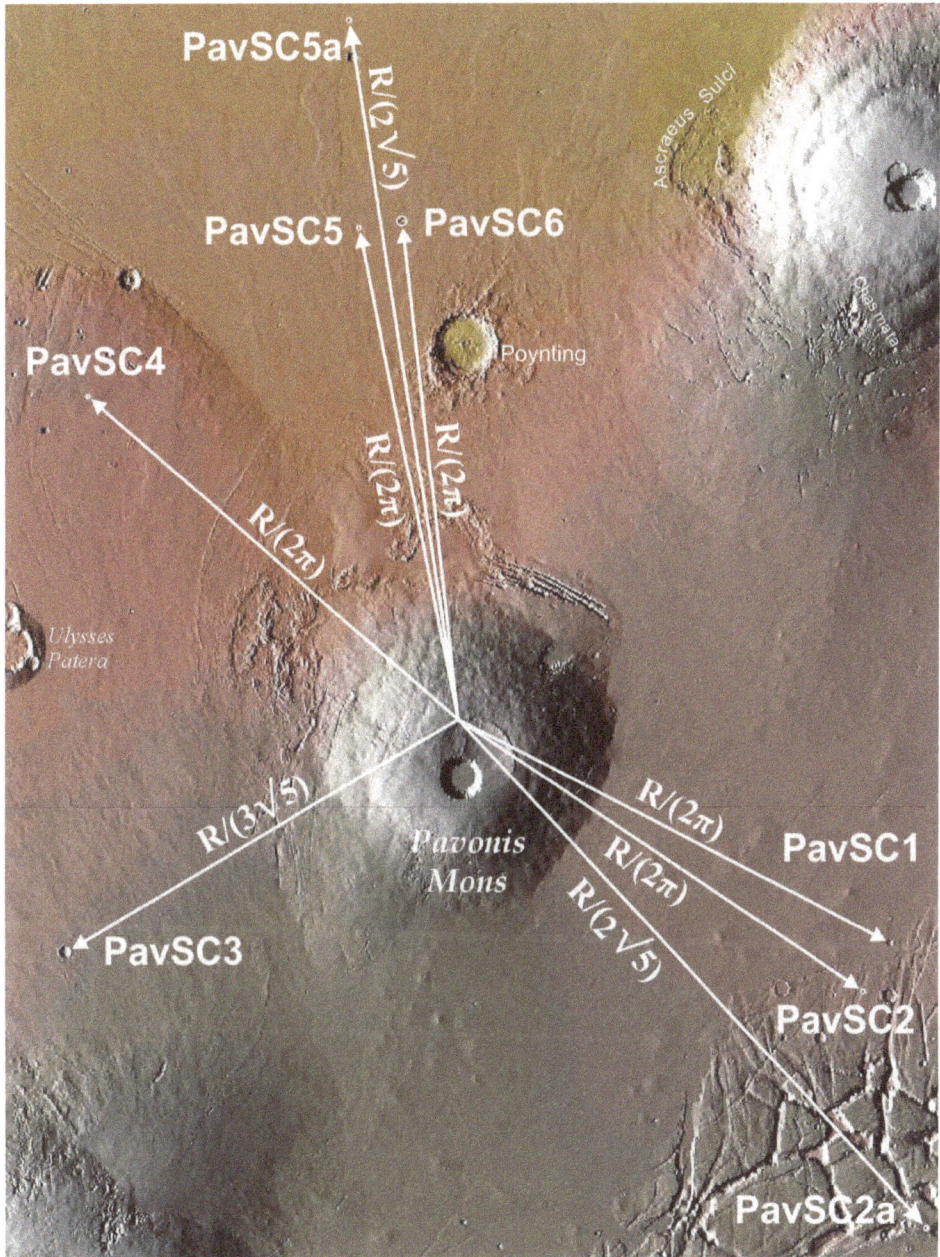

Fig. 5.2: *Survey crater model showing the 8 craters which were used for locating the survey centre for Pavonis Mons. This centre was determined by locating the coordinates of the point which gave the least squares differences for the sacred geometry distances of the 8 craters. The line to PavSC4 is within 7 min of e/π radians from due north in the counterclockwise direction. Also the line to PavSC3 is within 8 min of 120 degrees from due north in the counterclockwise direction. All craters except PavSC1 and PavSC3 have white lines drawn along their perimeters to enhance their visibility. USGS Astrogeology.*

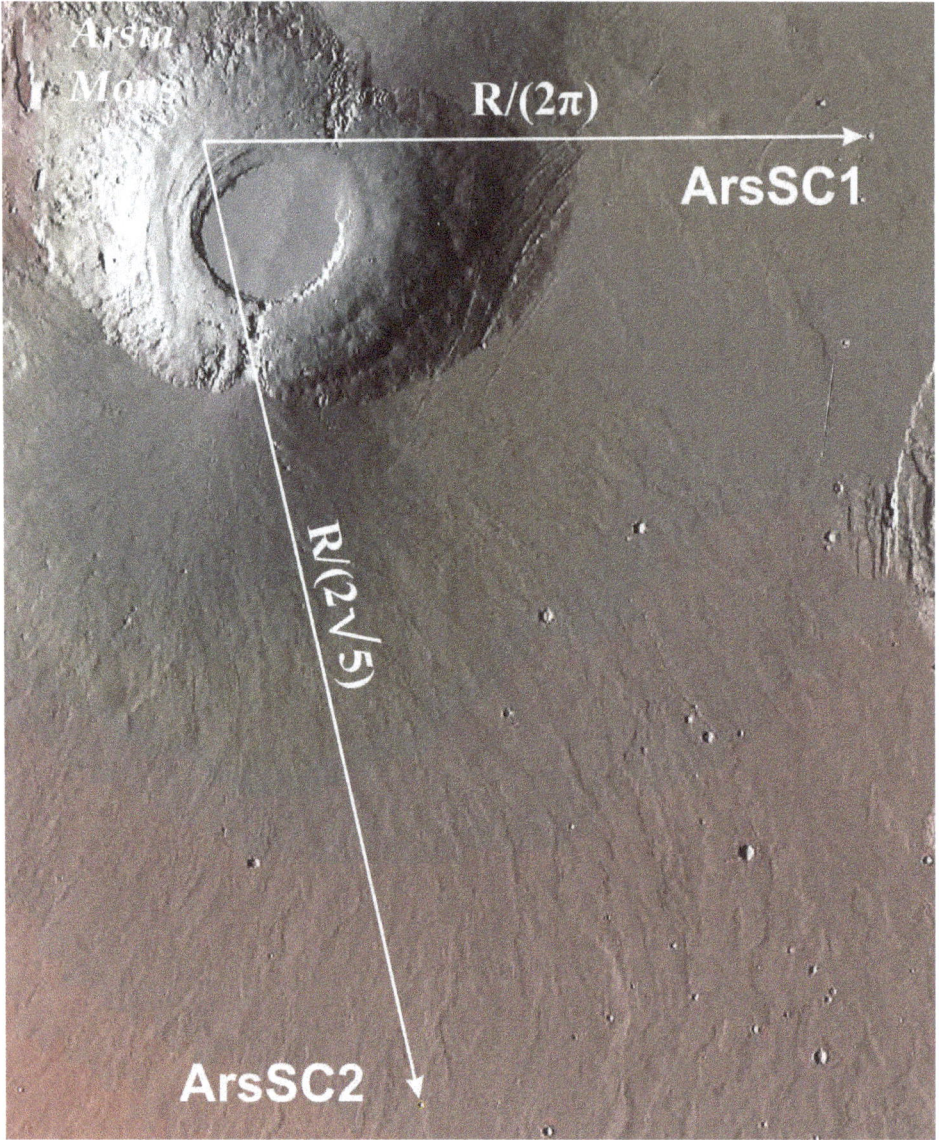

Fig. 5.3: *Survey crater model for locating the original centre of Arsia Mons. In this instance the 2 survey craters had different "sacred" distances. The survey centre for Arsia Mons was determined by locating the coordinates of the point which was R/(2π) km and R/(2√5) km from the crater centres of ArsSC1 and ArsSC2 respectively. USGS Astrogeology.*

(ArsSC1) being about 5.9 km in diameter and the southern crater (ArsSC2) about 5.6 km in diameter.

Coordinates (238.6754° E 8.0996° S) were then determined for the point on Arsia Mons which was exactly at the respective sacred distances from the 2 craters (Fig. 5.3). At first I was reluctant to accept the validity of these craters fulfilling the role of survey craters. What started to change my mind was the finding that the location they pointed to on Arsia Mons was almost exactly $R/\sqrt{5}$ km (1518.87 instead of 1518.82 km) away from the location determined to be the survey centre of Ascraeus Mons. But what really blew me away was discovering that the bearing of the rhumb line between the survey centres of these 2 mountains was calculated to be 40.8641 degrees which differs from the value (40.8682°) for $\mathrm{atan}(e/\pi)$° by only about 15 seconds of a degree[b]. This suggested that the bearing (clockwise angle from due north) of the base of the model isosceles triangle is extremely close, if not exactly equal to, $\mathrm{atan}(e/\pi)$ degrees.

As with Pavonis Mons, the survey centre of Arsia Mons did not lie within the boundaries of the current caldera. Instead it was positioned northwest of the main caldera and was just inside of a curved edge which could be construed as a trace of a more ancient caldera perimeter. It was also not far from the centre of an older version of Arsia Mons to the northwest of the main mass of the current mountain peak and therefore is a credible candidate for the centre of the original version of Arsia Mons.

Survey Craters for Olympus Mons

The next mountain to study was Olympus Mons. It did not take too long to discover 2 small craters that were, on average, about 430 km from the mountain "centre" determined from fitting a circle to the perimeter of the mountain. This value was reasonably close to 416.5 km which is $R/(3e)$ km. Hence, I calculated the coordinates of a point $R/(3e)$ km distant from each of these craters which I named OlySC1 and OlySC2. At a much later date, I discovered 2 other craters which were very close to being $R/(2\sqrt{5})$ km from this point. Since their bearing angles lay in between those of OlySC1 and OlySC2, I named the new pair as OlySC1a and OlySC1b so that I did not have to change the names of the first pair of craters. I recalculated the coordinates (227.3258° E 18.3570° N) of the survey centre for Olympus Mons by finding the location which gave the smallest sum of the squared differences between the actual distances and the theoretical distances for all 4 craters. This point was located within the present caldera of Olympus Mons just west of its eastern edge (Fig. 5.4).

Table 5.2 lists the coordinates and approximate diameters of the 4 craters used to determine the survey centre of Olympus Mons. The craters are all very similar in size, ranging from 6.2 to 6.6 km in diameter. The coordinates of the postulated original position of Olympus Mons given by the 4 survey craters turned out to be less than 1 km different than the distance $R/\sqrt{5}$ km from the survey point found for Pavonis Mons. Since this point is located within the present caldera of Olympus Mons it has a

Table 5.2: *Olympus Mons survey crater coordinates and distances from the survey centre.*

Crater Name	Diam- eter (km)	Latitude	Longitude	Theoretical Distance (km)		Actual Distance (km)	Differ- ence (km)
OlySC1	6.2	20.6218 ° N	234.3784 ° E	R/(3e) =	416.46	416.30	-0.16
OlySC1a	6.5	11.1733 ° N	238.3088 ° E	R/(2√5) =	759.41	759.62	0.21
OlySC1b	6.6	8.1201 ° N	235.2499 ° E	R/(2√5) =	759.41	759.36	-0.05
OlySC2	6.5	11.3938 ° N	226.3715 ° E	R/(3e) =	416.46	416.34	-0.12

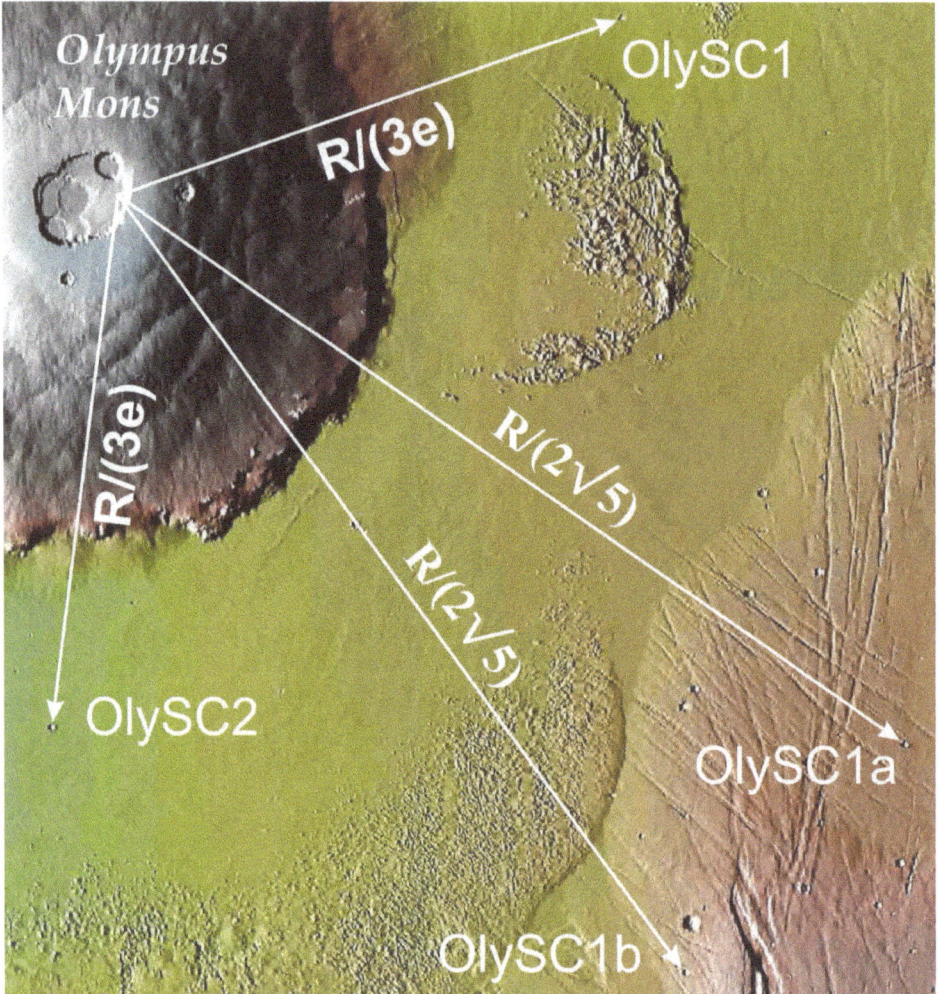

Fig. 5.4: *Survey crater model for Olympus Mons. OlySC1 and OlySC2 are at a sacred distance of R/(3e) km, and OlySC1a and OlySC1b at a sacred distance of R/(2√5) km from the survey centre of Olympus Mons. The craters are about 6.5 km in diameter and appear as tiny circles at the tips of the arrowheads. USGS Astrogeology.*

high degree of credibility for being the original centre of Olympus Mons.

Final Sacred Geometry Model for the Tharsis Giants

Now that the original mountain centres had been approximated from survey craters, it remained to be seen how well they fit the sacred geometry model from the previous chapter. In order to do this properly, however, a new dimension had to be added to this model, namely, that the base of the isosceles triangle would have a clockwise bearing of atan(e/π) degrees as suggested by the finding for the rhumb line joining the survey centres of Ascraeus Mons and Arsia Mons.

The sacred geometry model was constructed as a rhumb model as follows:

(1) The survey centre for Ascraeus Mons was selected as the anchor point for the model. The northeast end of the base of the isosceles triangle was positioned at this location. The coordinates of the southwest end of the base of the triangle were determined from a rhumb line with a clockwise bearing of atan(e/π) degrees and a length of R/√5 km from the survey centre of Ascraeus Mons. The base centre was determined from a rhumb line having a similar bearing angle but a length of R/(2√5) km instead of R/√5 km.

(2) The southeast end of a rhumb line with a clockwise bearing of atan(e/π) + 90 degrees and a length of R/√5 km was placed at the centre of the isosceles triangle base as obtained above. The coordinates of the line's northwest end marked the location of the apex of the isosceles triangle. The coordinates of all 3 vertices and centre of base of the model triangle were now available for distance calculations.

Table 5.3 lists the distances and bearing angles between the 4 mountains for 3 models: a model assuming a flat surface (linear model), the rhumb model, and the model obtained from simply joining the survey centres for the 4 mountains. Since the base and height of the rhumb model were set to precise distance and bearing angles, the rhumb model agrees perfectly with the linear model for all distances and bearing angles involving these components. However, because the rhumb model is placed on a curved surface, the sides between Olympus Mons and either Ascraeus Mons or Arsia Mons cannot be made equal to the linear model values. Neither can the 2 sides be made equal to each other since the side at higher latitudes has less distance to cover than the side at lower latitudes. The very shape of the triangle would have to be distorted in order to make them equal. Hence, the rhumb model distance between Arsia Mons and Olympus Mons was measured to be 4.60 km greater than

Mountain From	To	Formula	Linear Model Distance (km)	Bearing (degrees)	Rhumb Model Distance (km)	Bearing (degrees)	Survey Model Distance (km)	Bearing (degrees)	(Survey - Rhumb) Distance (km)	Bearing (degrees)
Arsia	Ascraeus	$R/\sqrt{5}$	1518.82	40.868	1518.82	40.869	1518.87	40.864	0.05	-0.005
Arsia	Pavonis	$R/(2\sqrt{5})$	759.41	40.868	759.41	40.869	760.72	40.911	1.31	0.042
Pavonis	Ascraeus	$R/(2\sqrt{5})$	759.41	40.868	759.41	40.869	758.15	40.818	-1.26	-0.051
Olympus	Pavonis	$R/\sqrt{5}$	1518.82	130.868	1518.82	130.869	1519.76	130.813	0.94	-0.056
Olympus	Arsia	$R/2$	1698.10	157.433	1702.70	157.064	1702.98	157.052	0.28	-0.012
Olympus	Ascraeus	$R/2$	1698.10	104.303	1668.16	104.566	1668.49	104.563	0.33	-0.003

Arsia Mons survey centre to rhumb model vertex — 0.13 km
Pavonis Mons survey centre to rhumb model base midpoint — 1.44 km
Ascraeus Mons survey centre to rhumb model vertex — 0.00 km
Olympus Mons survey centre to rhumb model vertex — 0.34 km

	Linear	Rhumb	Survey	Survey minus Rhumb
Angle (degrees) at Olympus Mons	53.130	52.498	52.489	-0.009
Angle (degrees) at Arsia Mons	63.435	63.805	63.812	0.007
Angle (degrees) at Ascraeus Mons	63.435	63.697	63.699	0.002

Table 5.3: *Model dimensions for Olympus Mons and the Tharsis Montes. The sacred geometry model from Chapter 4 set with a bearing angle of atan(e/π) degrees for the isosceles base is used for the linear and rhumb models. The linear model assumes a flat surface whereas the rhumb model is overlaid onto a spherical surface. The survey model is obtained from the mountain centre estimates calculated from survey craters. The last 2 columns give the differences in distance and bearing between rhumb and survey models. Also shown are the vertex and base midpoint differences between rhumb and survey models, and the differences between the 3 internal angles of the isosceles triangle for the rhumb and survey models. Bearings are in the clockwise direction.*

R/2 km, and the distance between Ascraeus Mons and Olympus Mons was measured to be 29.94 km shorter than R/2 km. The bearings of these sides were also significantly different from the linear model as were the internal angles at the base and apex of the isosceles triangle.

When we compare the rhumb model with the survey crater model in Table 5.3, it can be seen that the postulated architects chose to make the necessary distortions in exactly the same way as in the rhumb model. The survey centres for Olympus Mons, Arsia Mons and Pavonis Mons all lie within 1.44 km of the respective rhumb model triangle vertices and base midpoint. Distance measurements between the mountain survey centres (i.e., the sides and height line of the bisected triangle) are within 1.31 km of the corresponding distances of the rhumb model triangle. All of this of course suggests that we now have the correct model for the layout of the Tharsis giants. It is virtually locked into place not just by mountain survey centres but by the 16 survey craters which are all at "sacred" distances from the centres which they mark out.

The splendid fit of the final sacred geometry rhumb model to Olympus Mons and the 3 Tharsis Mounts can be seen in Fig. 5.5. It should be noted that despite the agreement of the linear model with the rhumb model in the distances for the height and base of the triangle, the linear model cannot be placed on the MOLA map since distance is nonlinear on this map. The length of the line marking the height of the triangle has to be slightly longer than the length of the line marking the base of the triangle to compensate for this nonlinearity which is due to the spherical nature of the planet. The rhumb model fits the mountain centres determined from survey craters so closely that they are covered over by the model lines. The perfection of fit is incredible when you take into account the vast distances which are covered by the geometric figure. The mountains are not only located precisely with respect to one another, but the base of the triangle is oriented so that its bearing is 40.8682 degrees from due north, the angle whose tangent is equal to e/π. The vertices at Olympus Mons and Ascraeus Mons both lie within currently existing calderas. The base midpoint is adjacent to the northern edge of an ancient caldera on Pavonis Mons. The vertex at Arsia Mons is just inside a very old caldera remnant and it marks out a centre of a faint outline of what looks like an ancient version of the mountain lying at the northwest perimeter of the current mountain. The apparent shift in calderas, especially for Arsia Mons, over billions of years may be the reason why it is so difficult for the modern astronomer to notice the pattern which now seems obvious. Also it is not credible in current thinking to allow for the possibility of artificiality in mountain placement, so the idea is never permitted to surface. It is hard not to see the inherent beauty of the geometric figure

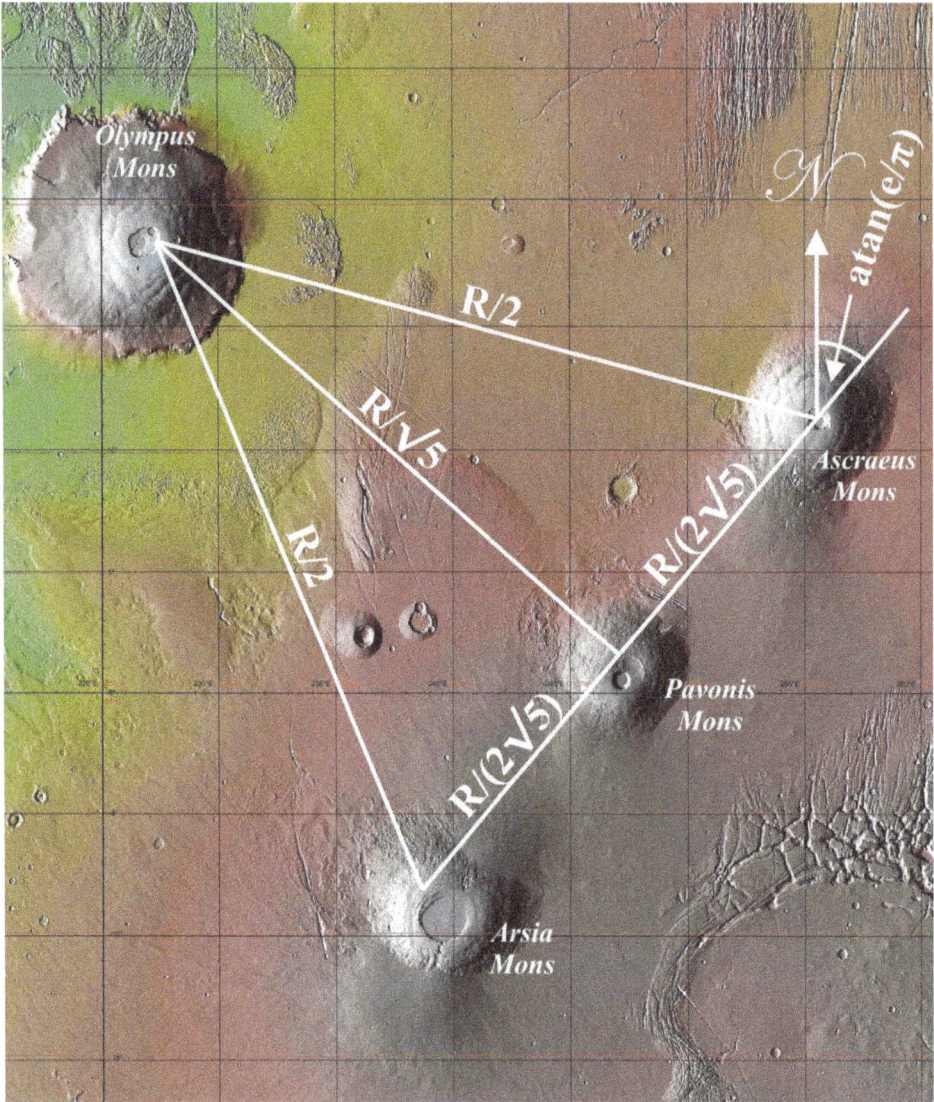

Fig. 5.5: *Final sacred geometry model for the Tharsis Montes and Olympus Mons. The model is drawn according to the rhumb model in Table 5.3. Since the vertices and mid-base of the triangle are all within 1.44 km of the survey centres for the various mountains, the model is virtually identical to that which would be obtained if lines were simply drawn between the survey centres. All component lengths of the bisected isosceles triangle which form the model incorporate the planetary radius plus an integer and/or the square root of 5. The base of the triangle is tilted at an angle of atan(e/π) = 40.8682° from due north in the clockwise direction. USGS Astrogeology.*

now that it has been uncovered.

The location of the survey centre of Pavonis Mons deviates slightly more than 1 km from the rhumb model. It also is slightly more than 1 km closer to Ascraeus Mons and more than 1 km further from Arsia Mons than $R/(2\sqrt{5})$ km. Notice that the largest errors in Table 5.3 are with lengths and bearing angles in which Pavonis Mons is one of the sites. Since some of the discrepancies are greater than my expectation of a ±1 km error in distance measurements, it is possible that the survey location of Pavonis Mons was not simply misplaced due to an error in my calculations but was deliberately placed slightly off the model to accommodate other measurements of sacred geometry involving this site. One such measurement is its latitude. It may have been shifted slightly northward to its value of 1.5993° N to be closer to a latitude of 1.6180° N so that it could pay homage to the value of the golden ratio φ. But why not make it more exact? I pondered on this for a long time when it suddenly struck me that if it was expressed in planetographic rather than planetocentric coordinates it just might coincide perfectly. I quickly found this to be true - my calculation of the survey centre's planetographic latitude was 1.6184° N which is only 1.4 seconds of a degree in error. There is only 1 other site which I discovered to have its latitude in terms of planetographic coordinates. It will be discussed in Chapter 9.

The symbolism of this "sacred" triangle seems to be centred around the double square whose diagonal is equal to $\sqrt{5}$ times its width. Each of the 2 equal sides of the isosceles triangle form the diagonal of a double square whose width is equal to 1/2 of the base and whose length is equal to the height of the isosceles triangle. The actual distances of the base and height of the triangle both incorporate the value $\sqrt{5}$, and what is not so obvious, the angles at the base for the linear model do so as well. These 63.435° angles are equal to $\mathrm{acos}(1/\sqrt{5})$, i.e., the angle whose cosine is equal to $1/\sqrt{5}$. They are also equal to $\mathrm{atan}(2)$ which probably honours the double square. In turn, the value $\sqrt{5}$ is the irrational component of the golden ratio and therefore represents that sacred number. The exactness of the formulae for the base angles is lost in the rhumb and survey models due to the curvature of the planet.

So in summary, the placement of Olympus Mons and the 3 Tharsis Montes can be modelled with a high degree of accuracy over enormous distances by a bisected isosceles triangle which incorporates sacred geometry for the equal sides, base halves and height. Each component is a measure of the planetary radius in some form or other, and each incorporates an integer and/or $\sqrt{5}$. The triangle is tilted to give the base a bearing of $\mathrm{atan}(e/\pi)$ degrees from due north. Furthermore, no less than 16 craters are at "sacred" distances from the various surveyed mountain centres. All of these distances incorporate a measure of the planetary equatorial radius together with either π, e or $\sqrt{5}$. For all this to happen simply by randomness is very difficult to believe.

Evaluation of Methods for Positioning Sites

With the high degree of closeness of fit of the surveyed mountain positions to the rhumb model in Table 5.3, we now have the capacity to evaluate the various methods available for positioning architectural sites on a spherical surface. If my theory about Martian architects positioning these sites is correct then they must have used a particular method to best convey their intended message. If a systematic method can be uncovered, this would greatly support my theory of artificiality of site placement. In Chapter 3 we saw that great circles were unlikely to be used because of the difficulty in representing a given bearing angle over the entire path between 2 sites. We can check out that assumption now by using great circles to determine the coordinates of the mountains using the same survey craters as were used with the rhumb line method. The results of this exercise are shown in Table 5.4. When the distances between the newly surveyed mountain positions were determined using great circles, it was found that they were virtually indistinguishable from those determined by the rhumb line method except for the distance between Olympus Mons and Ascraeus Mons which was 1.06 km shorter than for the rhumb line. The lack of a difference for most of the distances is due to the fact that these sites are close to the equator. The distance between Olympus Mons and Ascraeus Mons was shorter for the great circle than for the rhumb line since these 2 sites were located at a higher latitude and had a bearing angle quite deviant from a north - south direction. Now if we look at the mean bearings of the paths taken by the great circles, we find that they varied from the rhumb model by as much as half a degree. Thus the bearing of $atan(e/\pi)$ degrees is not adequately represented by great circle paths at all. If the mountains were positioned by architects, it is highly probably they would consider this to be a very important angle to portray. To do so, this result shows that they would have had to avoid using the great circle method for positioning the sites.

There is yet another method for determining distances between sites on a planet, and that is the method of using ellipses rather than great circles. This method is similar to using great circles except that it takes into account the fact that many planets like Mars and the Earth are oblate spheroids (a sphere squashed slightly at the poles) rather than perfect spheres. Thus the northern polar radius of Mars is 3376.20 km instead of the 3396.19 km found at the equator. The circumference of Mars has the shape of an ellipse when taken at any location other than at the equator, where it is circular. To take the ellipsoidal nature of Mars into account, I used Vincenty's formula[1] to calculate the positions of Olympus Mons and the Tharsis Montes utilizing the same survey craters as with rhumb line

Site-to-Site Distances (km)

Mountain		Theoretical		Rhumb Lines		Great Circles		Ellipses	
From	To	Formula	Distance	Distance	Deviation	Distance	Deviation	Distance	Deviation
Arsia	Ascraeus	R/√5	1518.82	1518.87	0.05	1518.86	0.04	1517.10	-1.72
Arsia	Pavonis	R/(2√5)	759.41	760.72	1.31	760.51	1.10	758.97	-0.44
Pavonis	Ascraeus	R/(2√5)	759.41	758.15	-1.26	758.36	-1.05	758.13	-1.28
Olympus	Pavonis	R/√5	1518.82	1519.76	0.94	1519.39	0.57	1519.39	0.57
Olympus	Arsia	~R/2	1702.70	1702.98	0.28	1702.97	0.27	1701.11	-1.59
Olympus	Ascraeus	~R/2	1668.16	1668.49	0.33	1667.43	-0.73	1667.15	-1.01

Site-to-Site Mean Bearings (degrees)

Mountain		Theoretical		Rhumb Lines		Great Circles		Ellipses	
From	To	Formula	Bearing	Bearing	Deviation	Bearing	Deviation	Bearing	Deviation
Arsia	Ascraeus	atan(e/π)	40.869	40.864	-0.005	41.342	0.473	41.405	0.536
Arsia	Pavonis	atan(e/π)	40.869	40.911	0.042	41.018	0.149	41.122	0.253
Pavonis	Ascraeus	atan(e/π)	40.869	40.818	-0.051	40.951	0.082	40.966	0.097
Olympus	Pavonis	atan(e/π)+90	130.869	130.813	-0.056	130.329	-0.540	130.322	-0.547
Olympus	Arsia		157.064	157.052	-0.012	156.610	-0.454	156.573	-0.491
Olympus	Ascraeus		104.566	104.563	-0.003	104.262	-0.304	104.270	-0.296

Table 5.4: *Comparison of methods for site positioning. Survey centres for mountain sites were determined by rhumb lines, great circles or ellipses using the same survey craters. Site-to-site distances and bearings (clockwise) were then calculated using the 3 different methods. Theoretical values are from the rhumb model in Table 5.3. Rhumb lines and great circles give the closest fit to theoretical distances. Great circles fail in mean bearing angles, and ellipses have the greatest deviations in distances and fail in mean bearings.*

positioning. It should be noted that the Vincenty formula required that the planetographic system of coordinates be used instead of the planetocentric since the planetographic system treats Mars as an ellipsoid rather than a sphere. I then used this technique to calculate the distances and mean bearings between the newly surveyed mountains. It can be seen from Table 5.4 in the ellipses columns that the distances are more than a kilometer shorter between Arsia Mons and either Ascraeus Mons or Pavonis Mons, and between Olympus Mons and Arsia Mons, than with the rhumb lines or great circles. Also the mean bearings of 5 out of 6 of the pathways of the ellipses between mountains deviate even more than with great circles, making the representation of atan(e/π) untenable. It is obvious from this that the geometry of ellipses would not have been used to position the sites or to portray the mountain pattern by my hypothetical architects. Only the rhumb model and rhumb line methodology work out properly for these mountains and their survey craters, and because of the consistent use of this technology, it is hard not to believe that the mountain sites were indeed artificially positioned.

Virtual Distance

The rhumb line methodology that I have been using assumes Mars to be a perfect sphere. At first this would seem to be an erroneous procedure since Mars is an oblate spheroid rather than a perfect sphere. The ellipsoidal nature of the planetary shape will thus lead to significant errors in distance measurements made with spherical geometry. Also high elevations above datum such as occur on the Tharsis Rise can lead to other serious measurement errors. Since the magnitude of the errors resulting from these anomalies could be in the order of several kilometers, they were a real concern to me. Yet the rhumb line technique worked perfectly for the geometric figure created by Olympus Mons and the Tharsis Montes, whereas the methodology of ellipses did not. I needed to find out why.

After giving considerable thought to this, I eventually reached the conclusion that for an observer located high above the surface of a planet, an architect wishing to convey the impression of a perfect geometric figure situated on an imperfect sphere would make the decision to position sites at the exact same coordinates which would be used if the sphere was perfect. To see the reasoning behind this conclusion, it is instructive to look at Fig. 5.6. When a planetary surface deviates from a perfect sphere, an architectural object can be either above, at, or below the elevation where the surface of the perfect sphere would pass. In Fig. 5.6, two sites are depicted with both sites lying on the actual planetary surface. The site

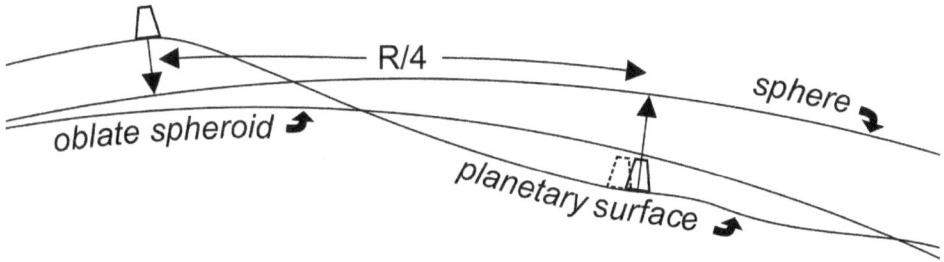

Fig. 5.6: *Positioning sites on a planetary surface to create a geometric shape. If an architect wishes to create the impression of the distance R/4 km from a viewpoint high above the planet's surface, the distance should be measured along the surface of an imaginary sphere of radius R km. If the measurement were made along the actual planetary surface, the lower site (dotted figure) in this instance would be placed too close to the upper site for the visual impression of an accurate distance.*

on the left lies above the perfect sphere's surface and the site on the right lies below it. Suppose the architect wanted the sites to be R/4 km apart where R is the equatorial radius of the planet. The architect could measure out the distance either along the actual surface of the planet or along the imaginary surface of a perfectly spherical planet of radius R km. If the former option was chosen, then the sites would appear to be closer to each other than R/4 km when viewed from outer space. If the latter option was selected, then the distance along the actual planetary surface would be longer than R/4 km. Which option would an architect choose? The main objective of an architect would be to correctly represent a geometric figure. The first option would present a distorted shape to a viewer high above the planetary surface. This would hardly seem to be very desirable. The only way to have the geometric figure appear to have the correct shape would be to position the sites according to distances measured along the imaginary spherical surface. If this option were chosen for the mountain and crater sites on Mars, then the intended distances between sites would be correctly estimated only by equations which assume a perfect sphere. From the close fitting results obtained with the bisected isosceles triangle model, it would appear that this is precisely what was done by my hypothetical Martian architects, and that the radius of the imaginary perfect sphere was chosen to be the equatorial radius R. With respect to positioning the sites according to the ellipsoidal surface of the planet, since this path is shorter than the spherical path, it would make the sites appear farther apart than with the virtual spherical path. This also is not desirable.

As a result of this analysis, I decided to continue with my practice of measuring site-to-site distances with rhumb line methodology which assumes a perfect sphere set to the equatorial radius of Mars. This decision was assisted by my choice of the planetocentric coordinate system which "assumes [the] planet is a sphere for map projections"[2]. Thus all of the distances presented in this book are actually virtual distances which suppose that all the sites are situated at the elevation of this virtual sphere. Distances measured by following the true surface of the planet would give a different result, but in my estimation, would not represent the postulated geometric figures accurately from the viewpoint of an observer high above the surface of the planet.

[a] *A radian is a unit of angular measure. One radian is equal to 180/π = 57.2958 degrees.*

[b] *The term 'atan' is used as an acronym for the arctangent function which returns the value of the angle (in degrees) whose tangent is the number in the brackets following 'atan'. Thus 45 degrees has a tangent of 1, and atan(1) is equal to 45 degrees. Similarly acos(x) returns the value of the angle whose cosine is x, and asin(y) returns the value of the angle whose sine is y. For those readers who would like to jog their memories of the terms tangent, sine and cosine: in a right angled triangle, take one of the angles which is not the right angle. Then the tangent of this angle is the number obtained by dividing the length of the side opposite to it by the length of the side adjacent to it. The sine of the angle is the side opposite to it divided by the hypotenuse. The cosine of the angle is the side adjacent to it divided by the hypotenuse.*

References

1. *The programming code which I constructed for calculating ellipsoidal distances and bearings was based on equations for Vincenty's formula as found in: Vincenty formula for distance between two Latitude/Longitude points. Chris Veness.*
   http://www.movable-type.co.uk/scripts/latlong-vincenty.html

2. *Letter from Mike Malin to colleagues concerning the decision to change the Mars cartographic coordinate system. June 20, 2001.*
   http://www.msss.com/mgcwg/mgm/letter.txt

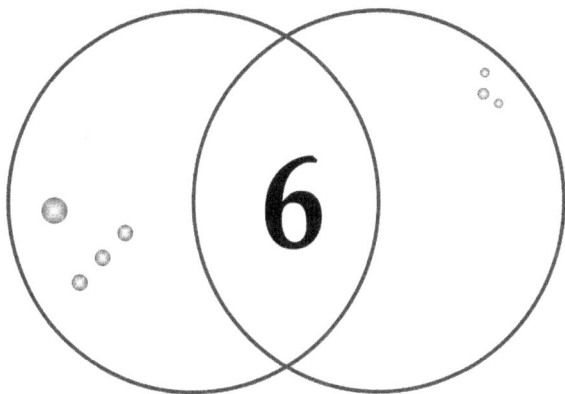

Elysium's Compass

W e now turn our attention to the next most impressive group of mountains, located west of the Tharsis giants about a quarter of the planet away (refer back to Fig. 2.2, left side). Here we see 3 mountains dominated by the 16 km high Elysium Mons. Just east of Elysium Mons are 2 smaller mountains each about 5 km in height. These are Albor Tholus to the south and Hecates Tholus to the north.

After seeing the remarkable pattern with Olympus Mons and the Tharsis Montes, the natural question to ask is whether there is a comparable design with the Elysium group. Albor Tholus and Hecates Tholus seem to lie on the same meridian, and so they could have been designed by our hypothetical architects to act as a pointer to the North Pole. But is there any proof of this, i.e., can the original mountain centres be determined with enough accuracy to test whether they do indeed lie on the same meridian? Also is there any sacred geometry present in the distances between these 3 mountains and/or a meaningful pattern to their original positions?

Survey Craters for Albor Tholus

To answer these questions, the first task was to try to determine whether any survey craters existed for these mountains such as was found for the mountains of the Tharsis Rise. As it turned out, the mountains on the

Elysium Rise were indeed marked out by a copious number of survey craters.

To start with, no less than 5 survey craters were found for Albor Tholus, all at a distance of approximately one fifteenth of the equatorial radius of Mars (R) from an estimate of the mountain centre determined as the centre of a circle fitted to its perimeter. A common point (150.3167° E 18.7937° N) was then determined to lie on Albor Tholus which was less than 1 km discrepancy from R/15 km from the centre of each of these craters. Their coordinates and deviations from the R/15 km distance are given in Table 6.1, and their positions are shown in Fig. 6.1.

The survey centre was found to be located just on the southern edge of

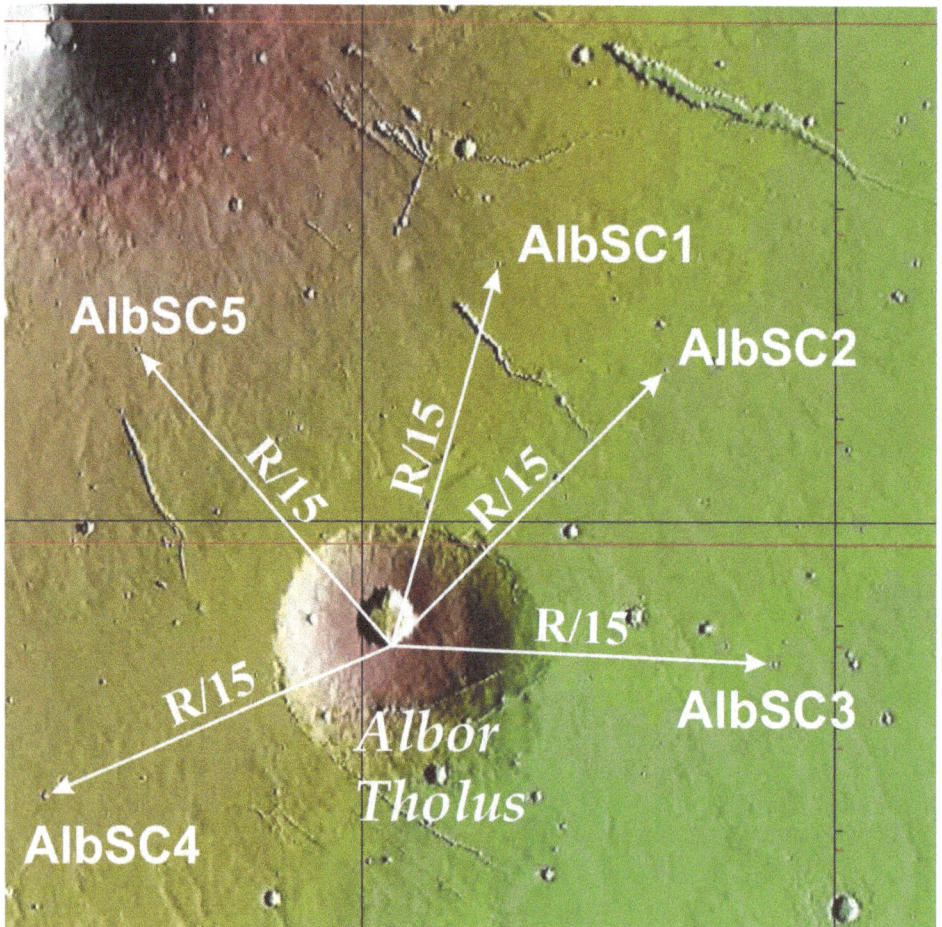

Fig. 6.1: *Survey craters used to locate the original centre of Albor Tholus. This centre was determined by calculating the coordinates of the unique point which was R/15 km from the centres of 5 craters, AlbSC1 to AlbSC5. Thin white lines have been drawn around the perimeters of AlbSC2 and AlbSC5 to enhance visibility of these craters. USGS Astrogeology.*

Table 6.1: *Survey crater coordinates for Albor Tholus and their distances from its survey centre.*

Crater Name	Diam-eter (km)	Latitude	Longitude	Theoretical Distance (km)	Actual Distance (km)	Differ-ence (km)
AlbSC1	2.5	22.4683 ° N	151.4441 ° E	R/15 = 226.41	226.61	0.20
AlbSC2	3.1	21.4623 ° N	153.2101 ° E	R/15 = 226.41	225.71	-0.70
AlbSC3	3.8	18.6267 ° N	154.3574 ° E	R/15 = 226.41	227.07	0.66
AlbSC4	5.7	17.3169 ° N	146.6128 ° E	R/15 = 226.41	226.34	-0.07
AlbSC5	3.7	21.6453 ° N	147.5971 ° E	R/15 = 226.41	226.82	0.41

the current caldera and lay only about 3 km south of the mountain "centre" determined by drawing a circle around the mountain using the western and northern outside edge of the mountain as a guide to fitting the circle. Hence, it was an excellent candidate for the original mountain centre. The angle between the lines joining AlbSC1 and AlbSC2 to the common point is 1.1 minutes of a degree less than φ/π radians (29.51 degrees) where φ is the golden ratio of 1.6180. The angle between AlbSC2 and AlbSC3 is only 37 seconds less than 47.00°.

The diameters of the survey craters for Albor Tholus ranged from 2.5 - 5.7 km. Those less than 4 km in diameter are the smallest survey craters observed so far.

Survey Craters for Elysium Mons

Moving northwest to Elysium Mons, a total of 7 craters were found which met the criteria for a survey crater. One of these craters was discovered much later than the other 6. To avoid having to rename the craters, I called the new crater ElySC2a as it lay between ElySC2 and ElySC3. This time, no less than 3 different "sacred" distances were present from a common location (coordinates = 147.1701° E 24.4910° N) on Elysium Mons which could be considered to be the original centre of this mountain. ElySC1 and ElySC2a were R/13 km from this point where R is the equatorial radius of Mars. ElySC2, ElySC3 and ElySC6 were all measured to be at a distance of R/(3π) km from the common point. The other 2 craters, ElySC4 and ElySC5, were at a distance measured to be R/(4π) km. These later 2 sacred distances are similar in format to R/(2π) km which was a very common sacred distance for the survey craters in the Tharsis group of mountains. The coordinates and deviations from theoretical distances of these craters are given in Table 6.2, and their locations are shown in Fig. 6.2.

Although the survey point seems to lie substantially off-centre of

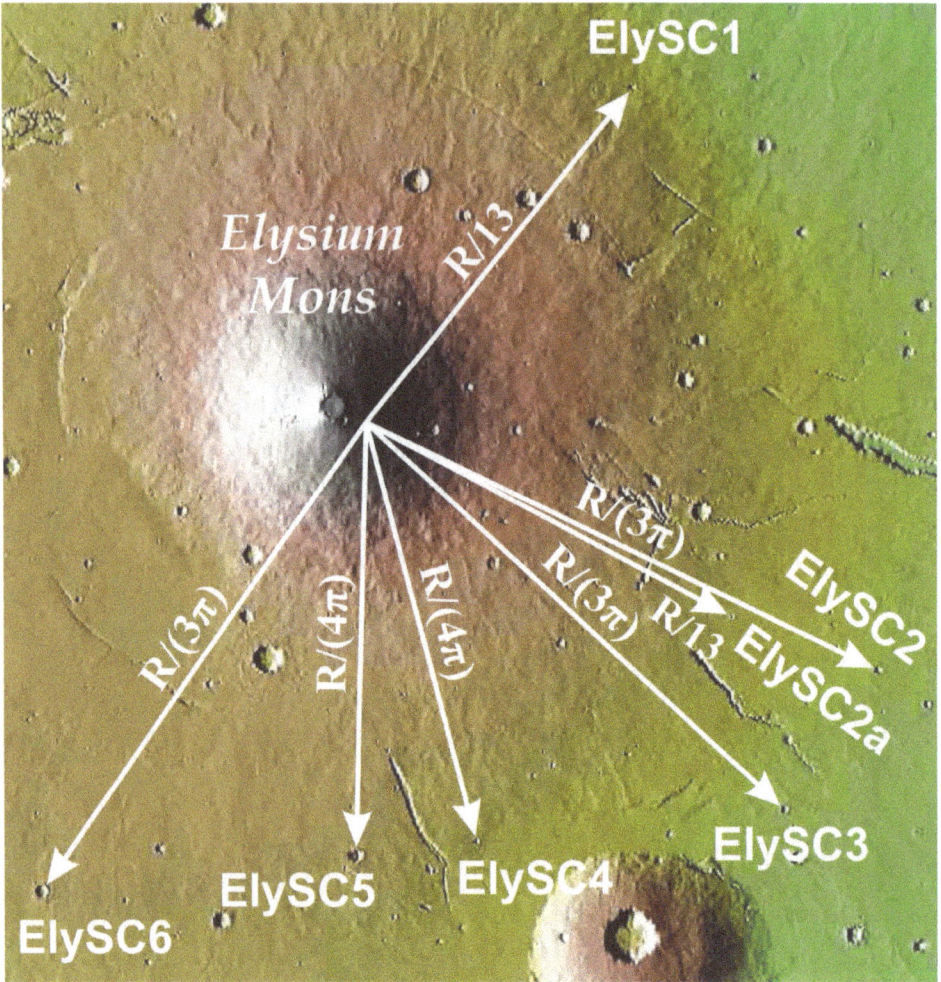

Fig. 6.2: *A total of 7 survey craters were used for locating the original centre of Elysium Mons. No less than 3 separate "sacred" distances were involved: R/13 km for ElySC1 and ElySC2a, R/(3π) km for ElySC2, ElySC3 and ElySC6, and R/(4π) km for ElySC4 and ElySC5. The bearing (clockwise) of the line to ElySC1 is within 5 min of atan(√5/e) degrees. The bearing (counterclockwise) of the line from ElySC3 to the Elysium Mons survey centre is only 1.0 min more than 48 degrees. The bearing (counterclockwise) of the line from ElySC4 to the Elysium Mons survey centre is only 54 sec less than asin(1/(e√2)) degrees. The ElySC2a crater has a thin white line drawn along its perimeter to enhance visibility. USGS Astrogeology.*

Elysium Mons, if a circle is drawn using the east side of the mountain as a guide for its circumference, the circle centre comes within 6 km of the common survey point. This would suggest that the centre of Elysium Mons has shifted to the west over the ages. The bearing of the line joining the common centre to ElySC1 is 4.8 min less than atan(√5/e) = 39.441

Table 6.2: *Elysium Mons survey crater coordinates and distances from the survey centre.*

Crater Name	Diam- eter (km)	Latitude	Longitude	Theoretical Distance (km)	Actual Distance (km)	Differ- ence (km)
ElySc1	5.0	27.8982 ° N	150.2855 ° E	R/13 = 261.23	261.21	-0.02
ElySc2	5.6	21.9063 ° N	153.1538 ° E	R/3π = 360.33	360.18	-0.15
ElySc2a	2.5	22.4683 ° N	151.4441 ° E	R/13 = 261.23	261.46	0.23
ElySc3	5.6	20.4281 ° N	152.0571 ° E	R/3π = 360.33	360.03	-0.30
ElySc4	5.3	20.0853 ° N	148.4519 ° E	R/4π = 270.25	270.44	0.19
ElySc5	10.7	19.9321 ° N	147.0403 ° E	R/4π = 270.25	270.32	0.07
ElySc6	9.1	19.5564 ° N	143.3452 ° E	R/3π = 360.33	360.13	-0.20

degrees in the clockwise direction from due north. If the bearings of ElySC3 and ElySC4 to, rather than from, the common centre are measured, it is found that the line from ElySC3 has a bearing angle (counterclockwise) only 1.0 min greater than 48.000 degrees. The bearing angle of the line from ElySC4 (counterclockwise) is only 54 sec less than asin(1/(e√2)) = 15.078 degrees.

ElySC2a is the same as AlbSC1. This is the first instance where a crater surveys 2 different mountains. It is the smallest survey crater, being only 2.5 km in diameter. The other 6 survey craters range in size from 5.0 - 10.7 km in diameter.

Survey Craters for Hecates Tholus

For Hecates Tholus, 4 survey craters were found with one pair (HecSC1 and HecSC4) at R/27 km from a common point (coordinates = 150.3210° E 31.5900° N) on the mountain, and the other pair (HecSC2 and HecSC3) at R/28 km from the same common point. The coordinates and deviations from theoretical distances of these craters are given in Table 6.3. The locations of the craters and the surveyed centre of Hecates Tholus are shown in Fig. 6.3.

Table 6.3: *Survey crater coordinates for Hecates Tholus and their distances from its survey centre.*

Crater Name	Diam- eter (km)	Latitude	Longitude	Theoretical Distance (km)	Actual Distance (km)	Differ- ence (km)
HecSC1	6.6	32.1655 ° N	152.7293 ° E	R/27 = 125.78	125.93	0.15
HecSC2	3.5	30.8346 ° N	152.5398 ° E	R/28 = 121.29	121.06	-0.23
HecSC3	5.2	29.6130 ° N	150.9379 ° E	R/28 = 121.29	121.34	0.05
HecSC4	6.6	33.3312 ° N	148.8866 ° E	R/27 = 125.78	125.69	-0.09

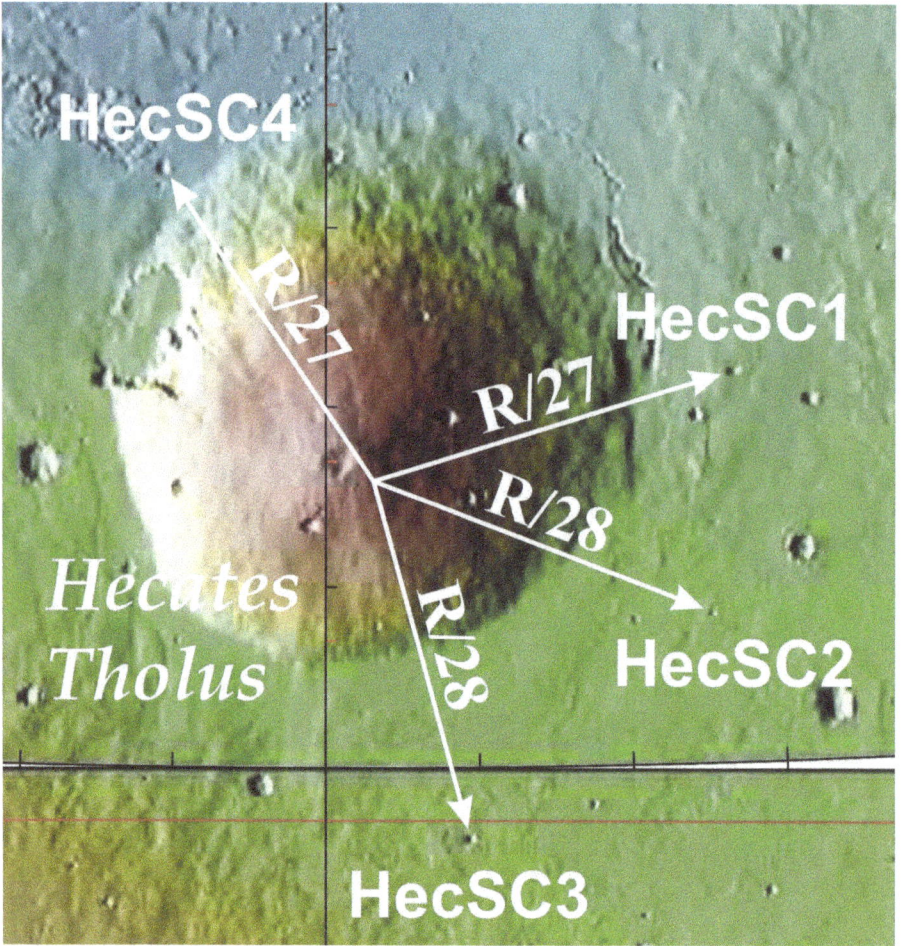

Fig. 6.3: *Survey craters used to locate the original centre of Hecates Tholus. The 4 survey craters had 2 different "sacred" distances: R/27 km for HecSC1 and HecSC4, and R/28 km for HecSC2 and HecSC3 where R is the equatorial radius of Mars. The bearing angle (counterclockwise) of the line from HecSC3 to the survey centre of Hecates Tholus is only 1.9 min greater than 15°. USGS Astrogeology.*

This common point is within 8 km of the southeast edge of the current caldera. Although it is off-centre for the main mass of the mountain it is very close to the centre of a circle created from the eastern portion of the southern perimeter of the mountain. What is most amazing about this location is that it has almost exactly the same longitude as for the survey centre of Albor Tholus, differing by only 15.5 seconds of a degree! This suggests that the postulated architects had designed the Albor Tholus and Hecates Tholus pair to be a marker of due north, acting like a huge compass. Such a marker would only be visible and useful from a high altitude, and therefore was probably intended for spacecraft approaching the planet for a landing.

An interesting feature about the survey craters for the Elysium group of mountains is that they are considerably closer to their respective mountains than those for the Tharsis group. The distance seems to reflect mostly the diameter of the mountain. Thus Elysium's survey craters are about 1.5 - 2 times closer than for the Tharsis group. The survey craters for Albor Tholus are more than twice as close and those for Hecates Tholus are about 4 times as close as for the Tharsis group. This only makes sense as you would not want the survey craters covered over by eruptions as the mountains grew over the ages. Hence, the survey craters for the larger mountains had to be placed further away than for the smaller ones.

Sacred Geometry Model For the Elysium Group of Mountains

Since there was no hypothesized model for the Elysium group of mountains other than that the longitudes of Albor Tholus and Hecates Tholus were the same, the only thing to do was to measure the distances between the surveyed mountain centres and look at the results to see if any patterns or "sacred" dimensions were present. The results from this exercise are presented in Fig. 6.4 and Table 6.4. Fig. 6.4 shows that an obtuse triangle is formed by connecting the survey centres of all 3 mountains together, with the obtuse angle at Elysium Mons and acute angles at Hecates Tholus and Albor Tholus. When the distances between the surveyed mountain centres were analysed, some startling findings emerged (Table 6.4). To begin with, the distance between Albor Tholus and Hecates Tholus was calculated to be only 0.91 km short of the length $R/(2\sqrt{5})$ km, the same distance between Arsia Mons and Pavonis Mons, and between Pavonis Mons and Ascraeus Mons. The next dimension of interest was the distance between Albor Tholus and Elysium Mons. This turned out to be only 0.14 km shorter that $R/(4\sqrt{5})$ km or 1/2 the theoretical distance between Albor Tholus and Hecates Tholus.

Hence, it seems that there was a very clear intent for the length of the above 2 sides by my postulated architects. Such was not the case for the side between Elysium Mons and Hecates Tholus. I now ran headlong into the problem of the existence of a multiplicity of possibilities for assigning a sacred geometry value for a given measured distance. In this instance, the sacred geometry value having the least deviation (0.62 km) from the measured value was found to be $\sqrt{3}R/13$ km. This formula did not fit into the numerical components used for the sacred distances of the other sides. There were also several clumsy formulae to choose from, all within 2 km of the measured distance. The only sacred formula which seemed to

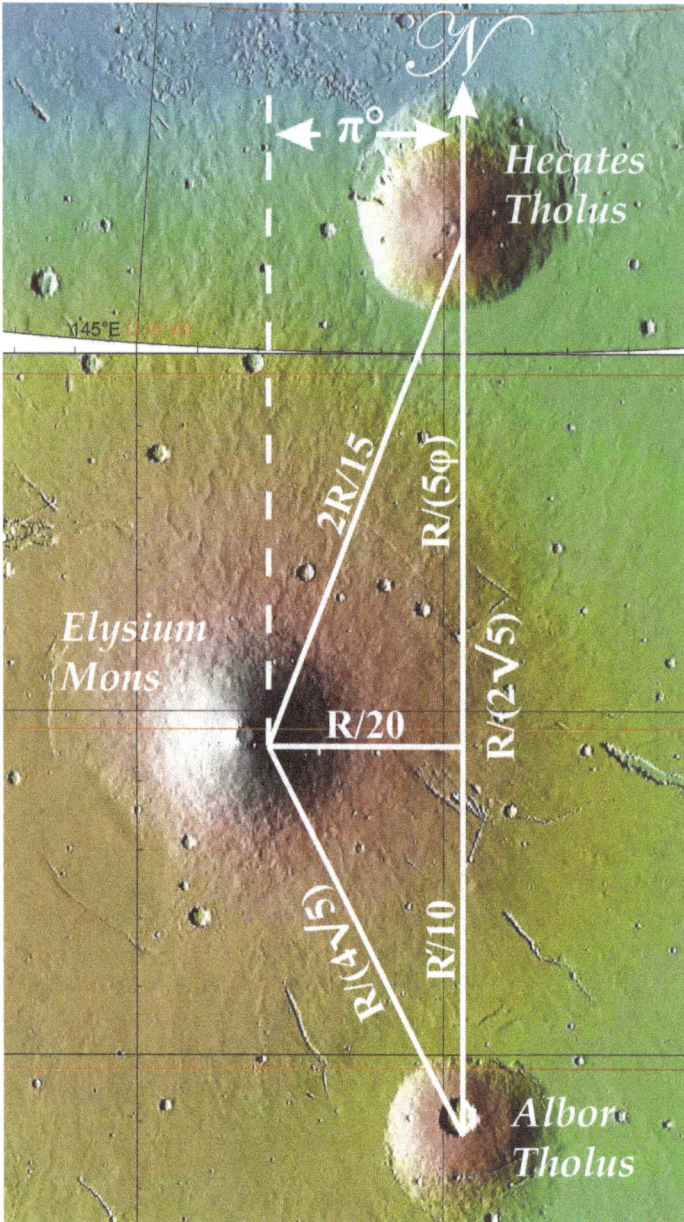

Fig. 6.4: *Triangular model for the Elysium group of mountains. The lines of the triangle join the "original" mountain centres determined from the survey craters shown in Figs. 6.1 - 6.3. The base of the triangle joining Albor Tholus to Hecates Tholus is close to R/(2√5) km, the same distance found for 1/2 the base of the bisected isosceles triangle used to fit the giant mountains of the Tharsis Rise. The longitude coordinates of Albor Tholus and Hecates Tholus are so similar that the bearing of the base can be assumed to be 0 degrees, and therefore points north - south. The side of the triangle which joins Albor Tholus to Elysium Mons is approximately equal to 1/2 of the base of the triangle. The longitude of Elysium Mons is very close to π° west of the triangle base. The height of the triangle would exactly equal R/20 km if Elysium Mons were at 24.5198° N which is only 1.7 minutes of a degree greater than the measured latitude of 24.4910° N. A higher latitude would give a value less than R/20 km, and a lower latitude, more than R/20 km. The length of the side joining Elysium Mons and Hecates Tholus could be approximated by 2R/15 km. The angle at Albor Tholus is only 1.3 min less than 10e degrees, and the angle at Hecates Tholus is 9.6 min more than 15√2 degrees. The angle of the main triangle at Elysium Mons is 3.4 min less than 93√2 degrees. USGS Astrogeology.*

Triangle Sides

From	To	Sacred Distance Formula	km	Survey (km)	Difference (km)	Bearing (degrees Counterclockwise)
Albor Tholus	Hecates Tholus	R/(2√5)	759.41	758.50	-0.91	0.000
Albor Tholus	Elysium Mons	R/(4√5)	379.71	379.56	-0.14	27.162
Hecates Tholus	Elysium Mons	2R/15	452.83	451.87	-0.96	158.627
Elysium Mons	Triangle base	R/20	169.81	169.85	0.04	-90.000
Albor Tholus	Height line	R'/10	337.62	337.71	0.09	0.000
Height Line	Hecates Tholus	R/(5φ)	419.79	420.79	1.00	0.000

Triangle Angles

Mountain	Linear (degrees)	Sacred angle Formula	Degrees	Survey (degrees)	Survey minus Sacred (degrees)
Albor Tholus	26.563	10e	27.183	27.162	-0.021
Hecates Tholus	22.023	15√2	21.213	21.373	0.160
Elysium Mons	131.414	93√2	131.522	131.465	-0.057

Table 6.4: *Sacred distance formulae were assigned to the rhumb line lengths between mountain survey centres for Elysium Mons, Hecates Tholus and Albor Tholus, and the intersect of the height line with the triangle base (see Fig. 6.4). Sacred geometry formulae were chosen on the basis of closeness to measured length, meaningfulness and simplicity. The difference column for distance refers to the differences between the survey centre measurements and theoretical values (sacred geometry formulae). Note that the counterclockwise bearing angle from Albor Tholus to Elysium Mons is the same as the angle at Albor Tholus. The 3 angles of the obtuse triangle are given for a flat surface model (linear), for closest fitting sacred geometry formulae, and for a spherical surface (survey) .*

serve the purpose was the sacred geometry value of 2R/15 km (deviation = 0.96 km), sacred because it contained the value of R, and also, because it had 5 as a factor in the denominator which, as we will see, is an extremely important number in Martian sacred geometry. In addition, the value 2R/15 is a relatively simple number. If we break this number into its factors we get 2R/(3x5) which can also be expressed as 2R/(3x√5x√5) which shows that √5 is actually present in all 3 sides of the Elysium group triangle. So, like the huge bisected isosceles triangle on the Tharsis Rise, √5 plays a pivotal role in the architecture of the Elysium group triangle.

This has brought me to the realization that smallness of deviation should not be the only criterion for assigning a sacred value to a distance or angle measurement. There are aesthetic considerations to be taken into account as well. And sometimes, in order to handle conflicting requirements for a model, some distortion or some slight inaccuracies have to be worked in. We already have evidence in the previous chapter that the "equal" sides of the bilateral isosceles triangle could not be made exactly equal on a spherical surface. Yet the intent clearly seemed to be to convey the impression of equality, and to represent the value of R/2 km for the diagonal of a double square for both halves of the bisected triangle. So all of this brought in some new perspectives to my quest to best interpret the supposed intentions of the hypothetical architects.

Getting back to the Elysium triangle, it occurred to me that the height of the triangle should be measured to see if any further information could be extracted. In order to do this, I simply used the latitude coordinate of Elysium Mons and the average longitude of the Albor Tholus and Hecates Tholus pair to create an imaginary point on the base of the triangle. The distance between this point and Elysium Mons would give the height of the obtuse triangle. The measured value was only 0.04 km less than R/20 km making this the obvious choice for a sacred distance. There were other sacred distance values which were less than 0.10 km from the measured value, but these were so clumsy that there was no problem in ignoring them. As will be discussed in Chapter 13, the shorter the distance, the more possibilities for sacred values exist.

Now there were only 2 more distance measurements to make, the distances between our imaginary point on the base of the triangle to Hecates Tholus and to Albor Tholus. Once again, there were several possible sacred distance values for each. The simplest sacred distance that could be applied to the distance to Albor Tholus was R/10 km. However, its deviation was a rather large 1.91 km. I then noticed that if you used the northern polar radius (see Chapter 9) instead of the equatorial radius of Mars, the deviation was reduced to a mere 0.09 km! This was so elegant that I decided that this was probably the intent of the postulated architects. I labelled the polar radius as R' in Fig. 6.4 to distinguish it from the equatorial radius R. It was a very simple formula, and it gives the impression that this side is exactly twice the height of the main triangle even though the radii are different for both sides. And what was starting to become apparent, the southern right triangle created by the height line mimics a half of the bisected isosceles triangle of the Tharsis Montes and Olympus Mons. It represents 1/2 of a double square, but this time, the slight distortion required for fitting on a spherical surface was applied to the height (distance from the imaginary point to Albor Tholus) rather than to the diagonal.

Two reasonable candidates for sacred distance values could also be applied to the distance between the height line of the main triangle and Hecates Tholus. These were $(R/7)(e/\pi)$ km (deviation = 0.995 km) and $R/(5\varphi)$ km (deviation = 0.999 km). As the deviations were very close to each other this did not factor into my choice. I decided to use the latter value for the label on Fig. 6.4 due to its greater simplicity although it is likely that the postulated architects used both to convey more than 1 meaning. The former pays homage to the bearing angle of the base of the bisected isosceles triangle of Chapter 5, and the latter strongly refers to the pentagram which has 5 star points and dimensions from which φ can easily be obtained (see Chapter 9).

I decided next to test out my choice of sacred distances for side lengths by placing the triangle on a flat surface to allow me to apply 2-dimensional geometry. Since there were 2 right triangles created by the height line, the Pythagorean theorem of the sum of squares could be used to calculate the hypotenuse in each triangle. If we ignore the use of the polar radius for now, the sum of the squares is $(R/20)^2 + (R/10)^2 = 5R^2/400$ for the southern triangle. The square root of this number works out evenly to $R/(4\sqrt{5})$ which is exactly the number which was found to be the sacred distance between Albor Tholus and Elysium Mons. Unfortunately, the square root of the Pythagorean sum of squares does not cancel down to a simple sacred number for the northern right angle triangle. However, the sacred number which comes closest to the numerical value of the hypotenuse is only 0.01 km shorter. And guess what that sacred number is? None other than 2R/15 km! This gives strong confirmation that my choices were likely to have been what the postulated architects originally intended. Depending on the symbolism or concept that was intended to be conveyed, either the polar radius or the equatorial radius could be used interchangeably for the distance between Albor Tholus and the intersect of the height line with the base of the main triangle.

The angle at Albor Tholus (fitted triangle, Fig. 6.4) was measured to be

27.1616 degrees which is only 1.3 min of a degree smaller than 27.1828 degrees, the value of 10e. The angle at Hecates Tholus was measured to be 21.3737 degrees which is 9.6 minutes more than $15\sqrt{2} = 21.2132$ degrees. Finally, the angle at Elysium Mons was determined to be 131.4651 degrees which is 3.4 minutes less than $93\sqrt{2} = 131.5219$ degrees. Angles were also calculated with the triangle being placed on a flat surface (column labelled 'linear' in Table 6.4) using the assigned sacred values for the sides of the obtuse triangle. The angle at Elysium Mons showed the least difference between the flat surface and the spherical surface angles.

Besides the discovery of the virtually exact alignment of Albor Tholus and Hecates Tholus along the same meridian line, another truly astonishing finding was that the difference between the longitude of Elysium Mons and the mean longitude of Albor Tholus and Hecates Tholus was 3.1488 degrees. This is only 25.8 sec different than the value for π (3.1416). Since the distance between meridian lines decreases as one moves from the equator to the poles, there is only one latitude at which the difference between the meridian lines passing through Elysium Mons and Albor/Hecates Tholus will be equal to R/20=169.81 km, and that latitude is extremely close to the latitude of Elysium Mons itself where I measured π degrees of longitude to be 169.46 km. In other words, if Elysium Mons had been placed at even a slightly higher or lower latitude, it would have been impossible to have the height of the triangle equal to both R/20 km and $\pi°$ of longitude. This is yet another indication that the placement of these mountains was exceedingly well thought out.

The sheer number of survey craters (a grand total of 16) present for the 3 mountains of the Elysium group suggests that it was very important to locate the position of these mountains accurately. The size of the survey craters ranged from 2.5 – 10.7 km in diameter. Remarkably, in the distances and the angles between the mountains, this single triangle manages to incorporate R, φ, e, $\sqrt{2}$, $\sqrt{5}$, and the whole numbers 2, 4, 5, 10, 15 and 20. It also manages to exactly point to the North Pole and to mark out π degrees of longitude at the same time. This is at least as ingenious as the sacred geometry triangle of Tharsis! All by a chance of nature?

While the survey centre for Elysium Mons appears to be placed off-centre towards the east, more evidence will be uncovered later in this book and in the 2nd book of the *Intelligent Mars* series that the location of its longitude coordinate creates very important longitude displacements to other major sites on the planet. This is also true of the survey centre for Pavonis Mons which, like the survey centre for Elysium Mons, was marked by a very large number of survey craters, and was slightly displaced from the sacred geometry model which fits Olympus Mons and the other 2 Tharsis Montes so well.

7

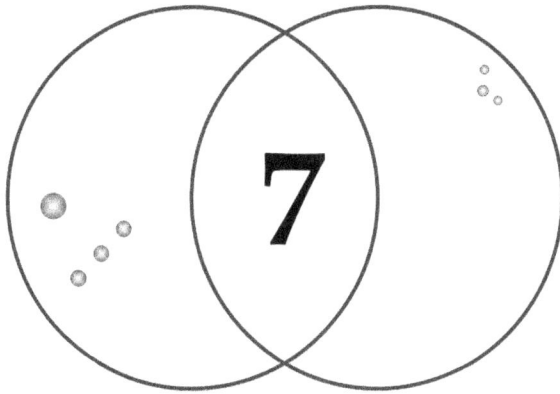

Lesser Mountains Part I. Mountain Groups

With Olympus Mons, the Tharsis Montes and the Elysium group, we have covered the tallest of the Martian mountains. But there are many more mountains yet to be looked at on Mars, and although lower than the Tharsis giants, some would still be considered to be very large by terrestrial standards. For instance, there is the low but massive Alba Mons which covers a vast area on the northern edge of the Tharsis Rise. There also are several more mountains on the Tharsis Rise, especially to the northeast of Ascraeus Mons. To the southeast of Elysium Mons is Apollinaris Mons. Although the most prominent mountains are mostly found in the northern hemisphere, there exists a large number of mountains in the southern hemisphere as well. In all, the USGS Gazetteer of Planetary Nomenclature (http://planetarynames.wr.usgs.gov/Page/MARS/target) lists more than 70 unique montes, mons and tholi for Mars. However, in order to keep the data in this book to manageable levels, I decided to limit myself to the area between Elysium Mons to the west, the Fesenkov Crater to the east, Arsia Mons to the south and Alba Mons to the north (see Fig. 7.1). In this vast area which encompasses more than one-third of the planet in terms of longitude and over one quarter in terms of latitude, I will discuss an additional 11 mountains and 5 major craters which have not yet been analysed in previous chapters.

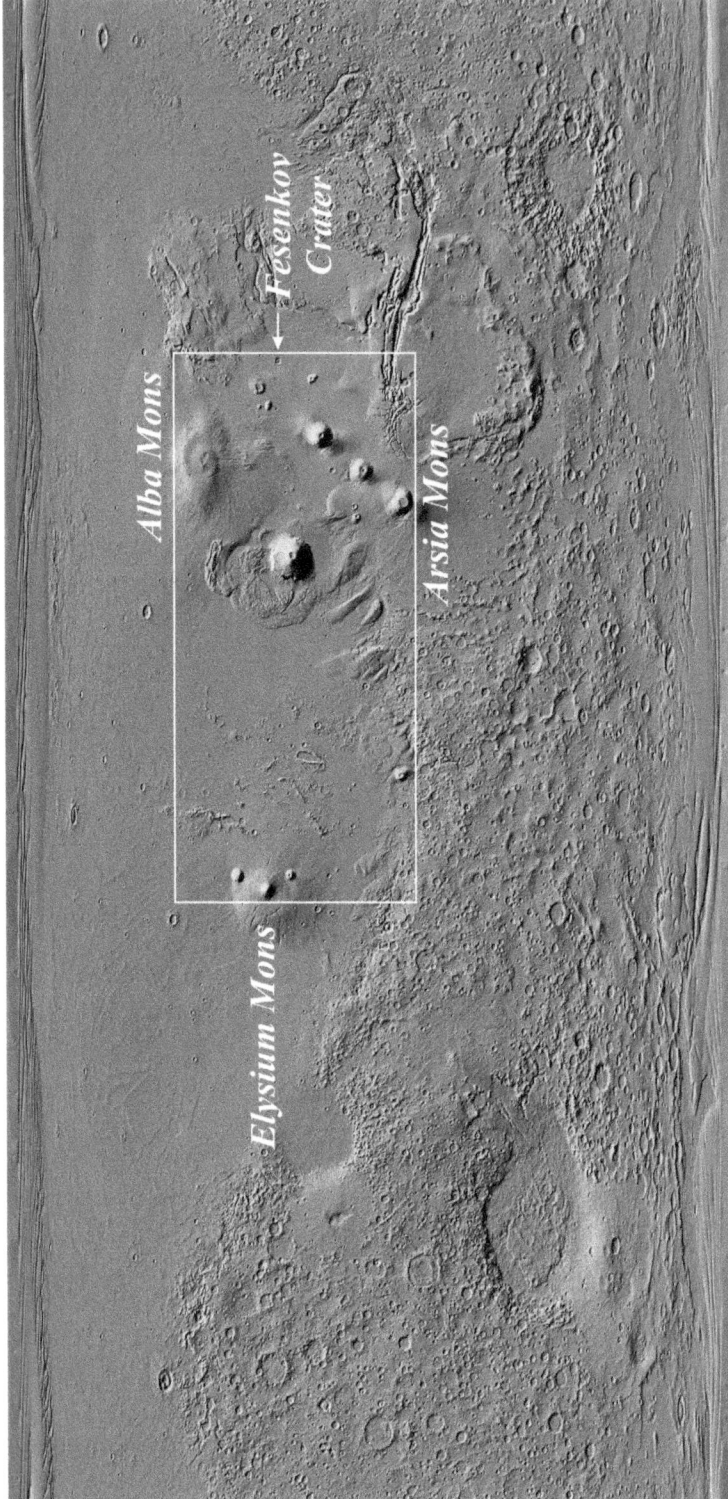

Fig. 7.1: *Area examined in this book for mountain and crater patterns (white rectangle). The area is bounded by Elysium Mons to the west, Alba Mons to the north, the Fesenkov Crater to the east, and Arsia Mons to the south. This map shows the entire surface of Mars. MOLA Science Team Courtesy NASA/JPL-Caltech.*

Because there were so many mountains that needed to be covered in this area, I decided to break up their analysis into 2 parts, giving each their own chapter. The first part deals with mountain groups: (a) Biblis Tholus and Ulysses Tholus, (b) Uranius Mons, Ceraunius Tholus and Uranius Tholus, and (c) Tharsis Tholus which has 2 peaks that I will treat as 2 separate mountains. The second part deals with single mountains scattered over a vast area: Apollinaris Mons, Alba Mons, Issedon Tholus and Jovis Tholus which has 2 centres.

With the 2 sacred geometry models uncovered so far, the tendency is to think that this is as far as it goes. Surely all the other mountains must simply be examples of your regular brand of mountain, each having sprung up at randomly positioned weak spots in the Martian crust in order to relieve pressure buildup in the planet's interior. Like everyone else I wanted to believe that Nature on Mars behaves the same way that Nature behaves on earth. I thought that perhaps there might be a couple more examples of planned positioning. For instance, it looked like the location of Alba Mons with respect to Olympus Mons and Ascraeus Mons might form the elusive equilateral triangle that I had originally been looking for. I had also previously looked for an equilateral triangle between Ceraunius Tholus, Tharsis Tholus and Ascraeus Mons, but came up empty handed. So overall, I did not have too many great hopes, and I thought that I should be able to wrap things up rather quickly from here. I could not have been more wrong, as you shall soon see.

I have discovered that there are an enormous number of sacred geometry triangles and distances involving the lesser mountains of Mars and several important craters! This chapter and the following one will allow you to get a firmer grasp on these other mountains and craters which might easily be overlooked. You will have the opportunity to see several instances of sheer genius in the patterns and relationships created by the layout of these structures. And later in Chapter 8 you will also learn about the mountain that just might be a giant pyramid – a pyramid that totally miniaturizes the great pyramids of Egypt and even the D&M pyramid in the Cydonia region of Mars which has received much attention in recent years.

Biblis Tholus and Ulysses Tholus

I shall start with Biblis Tholus and Ulysses Tholus, situated just west of the Tharsis Montes. Located west of Pavonis Mons and north of Arsia Mons, the elevation of Biblis Tholus is 7,198 meters above datum and that of Ulysses Tholus is 5,863 meters[1]. Biblis Tholus is about 170 km long by about 100 km wide and rises about 3 km above the surrounding terrain. It has a caldera which is 53 km in diameter and 4.5 km deep[2]. Ulysses Tholus has a diameter of about 100 km with a caldera diameter of about 56 km.

My initial measurements showed that Biblis Tholus was about equidistant from Pavonis Mons and Arsia Mons. Eventually I found a location on Biblis Tholus (see Fig. 7.2) that exactly fit both distances to the sacred geometry number $R/(3\sqrt{3})$ which is equal to 653.60 km. The coordinates of this point are 236.1176° E and 2.6278° N, and it sits just inside the northern edge of the current caldera. This is definitely a plausible location for the original position of the centre of this mountain, and ultimately its position is based on the survey centres of Pavonis Mons and Arsia Mons rather than any survey craters of its own. It lies about 14 km from the centre of a circle drawn around the perimeter of the southern portion of the mountain. However, the centre of a circle constructed from the perimeter along the northeast edge of the mountain is only about 1 km from the surveyed location. With this location then, Biblis Tholus forms a sacred geometric isosceles triangle with the base being the line between the survey centres of Pavonis Mons and Arsia Mons.

I next found a location on Ulysses Tholus (Fig. 7.2) that was exactly the same distance away from Arsia Mons as Biblis Tholus [$R/(3\sqrt{3})$ km], and was also exactly $\sqrt{3}R/(7\varphi)$ = 519.36 km from Pavonis Mons. The coordinates of this point are 238.4356° E and 2.9244° N, and it lies only about 1 km from the centre of a circle drawn around the perimeter of the Ulysses mountain. Once again, we have a surveyed location based on the surveyed centres of mountains rather than on survey craters. And once again, we discover a sacred isosceles triangle, this time with its apex at Arsia Mons and its base between Biblis Tholus and Ulysses Tholus.

The choice of a sacred distance between the survey centres of Biblis Tholus and Ulysses Tholus presented some difficulties. Like other short distances, there are many possible alternative sacred distances. The one that intrigued me the most was $e\varphi R/108$ = 138.31 km. At first, the denominator seemed rather clumsy, but then I realized that this number is the size (in degrees) of the inner angle of a regular pentagon or the angle between the star points of a pentagram, both geometric designs in which the numbers 5 and φ play prominent roles. It differed only 0.05 km from the measured distance of 138.36 km.

An alternative sacred distance between Ulysses Tholus and Pavonis Mons is $eR/(11\varphi)$ = 518.69 km. This is only 0.67 km from the centre produced with the $\sqrt{3}R/(7\varphi)$ km distance. An acceptable alternative sacred distance between Ulysses Tholus and Biblis Tholus is $\sqrt{3}R/(30\sqrt{2})$ = 138.65 km. Although the best overall fit was given with the distances selected in the previous 2 paragraphs, it is possible that all of the sacred

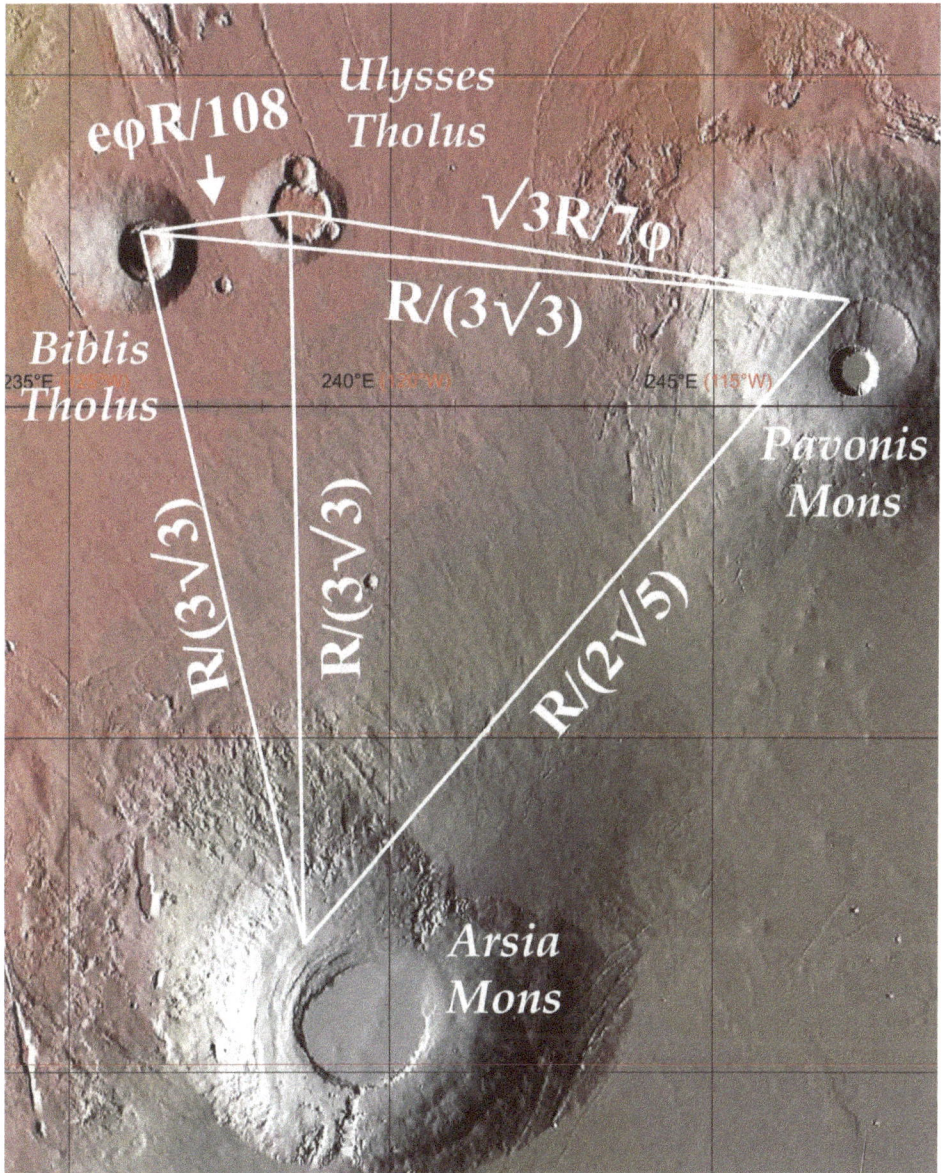

Fig. 7.2: *Sacred geometry patterns for Biblis Tholus and Ulysses Tholus. These 2 mountains together with Pavonis Mons and Arsia Mons create 4 triangles, 2 of which are isosceles using R/(3√3) km for their equal sides. The side between Arsia Mons and Biblis Tholus is common to both isosceles triangles. They seem to celebrate primarily the value of √3 although the golden mean is also present in the forms of φ and √5. USGS Astrogeology.*

distance alternatives were intended to be represented by this same geometric configuration in order to celebrate different symmetries and sacred numbers.

So here we have 2 sacred geometric isosceles triangles, one with its apex at Biblis Tholus and the other with its apex at Arsia Mons. The 2 equal sides of each of these 2 triangles have exactly the same value, namely R/(3√3) km, with the side between Biblis Tholus and Arsia Mons being common to both triangles. For what seems to be a celebration of √3, another important irrational number in sacred geometry, the most likely sacred distance from Ulysses Tholus to Pavonis Mons is equal to √3R/(7φ) km and √3 can also be found in the alternative distance of √3R/(30√2) km between Ulysses Tholus and Biblis Tholus. The finding of such symmetry not only left me with a great sense of awe but also suggested that I was on the right track.

The Eastern Mountains

A cluster of 3 mountains lies to the northeast of Ascraeus Mons while almost due east of Ascraeus Mons there is situated a pair of volcanic peaks collectively known as Tharsis Tholus. The question that naturally arose from my past experience with the giant Martian mountains was whether or not there was some sacred geometric pattern to these smaller mountains. Also, were they related to the Tharsis Montes and Olympus Mons in any way? After many preliminary measurements of distances between mountains and between mountains and prominent craters, I decided to try to fit all of these eastern mountains to sacred distances from 2 large craters, the Paros Crater (261.8686° E 21.9949° N) and the Fesenkov Crater (273.4705° E 21.6321° N), which were close by. With 2 craters, I could triangulate in an attempt to obtain the coordinates of the original centres of the mountains. If these craters were used by the postulated architects of mountain placement as survey craters, I would be able to find sacred distances to credible centres of the mountains. If I was on the wrong track, it would become obvious in very short order.

Ceraunius Tholus, Uranius Tholus, and Uranius Mons

The most obvious mountain to test the suitability of the Fesenkov and Paros craters as survey craters was Uranius Mons since it was located more centrally between the 2 craters. This mountain is the most easterly of the group of 3 mountains to the northeast of Ascraeus Mons. Its elevation is estimated to be 4,853 meters above datum[1], but is only about 1,350 meters above the northern plains[3]. I was not disappointed in the fitting process since I was able to find sacred numbers for the distances between a credible centre for this mountain and the 2 craters (Fig. 7.3). The surveyed centre for Uranius Mons had the coordinates 267.0163° E

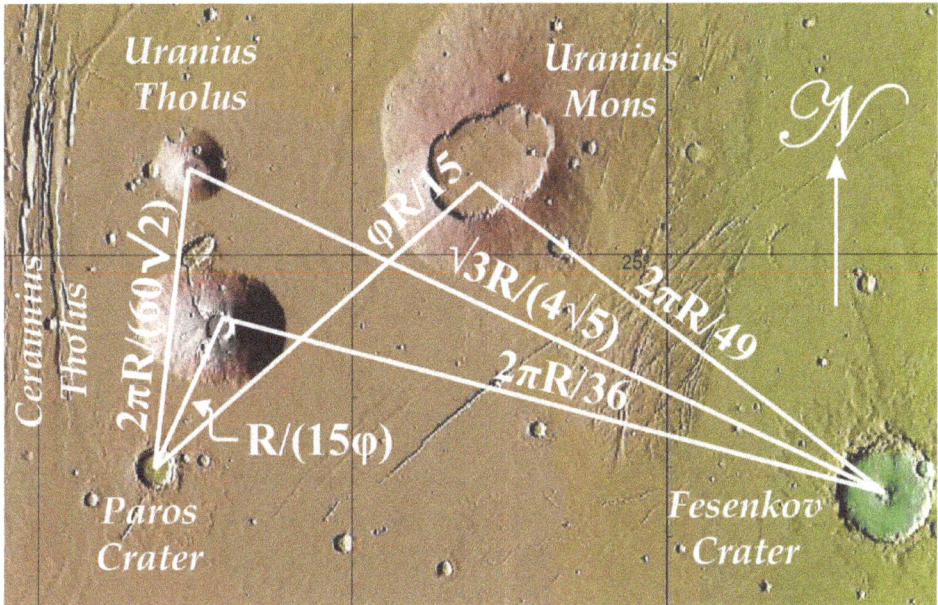

Fig. 7.3: *Determination of mountain centres for Ceraunius Tholus, Uranius Tholus and Uranius Mons. All mountain centres were surveyed from the Paros and Fesenkov Craters. Note the symmetry in 15 and φ for the Paros survey distances for Ceraunius Tholus and Uranius Mons. Note also the use of the equatorial circumference of Mars in the sacred distances for Ceraunius Tholus and Uranius Mons from the Fesenkov crater, as well as in the sacred distance from the Paros Crater to Uranius Tholus. USGS Astrogeology.*

26.0069° N and was placed $2\pi R/49$ = 435.49 km from the Fesenkov Crater and $\varphi R/15$ = 366.34 km from the Paros Crater (where φ is the golden mean of 1.6180). While these 2 numbers do not make an obvious pair, and the number 49 seems excessive (but see below), they nevertheless marked out a centre for Uranius Mons which was virtually identical to the centre of a circle which followed the southwestern perimeter of the mountain.

I next moved on to Ceraunius Tholus. This mountain is estimated to be 2 - 3 km high and about 3 billion years old[4]. From an examination of the USGS Astrogeology map, it appears to have 3 distinct circular perimeters and thus must have wandered considerably in location mostly in an east-west direction during its evolution. Once again I used the Fesenkov and Paros craters to survey out a credible mountain centre based on sacred numbers even though Ceraunius Tholus is much closer to the Paros Crater than to the Fesenkov Crater. Once again I was successful, finding a survey centre (coordinates 262.9598° E 24.1315° N) that was $2\pi R/36$ or 592.75 km from the Fesenkov Crater and $R/(15\varphi)$ or 139.93 km from the Paros Crater. Do you notice the symmetry with the distances for Uranius

Mons? Both mountains have the planet's equatorial circumference in their distances from the Fesenkov Crater and both use the numbers φ and 15 for their distances from the Paros Crater! And the sacred distance from the Fesenkov Crater to Ceraunius Tholus has the square of 6 in its denominator, and the sacred distance from the Fesenkov Crater to Uranius Mons has the square of 7. Once again we see a work of genius. The survey centre for Ceraunius Tholus was found to be very close to that for the centre of a circle which followed the perimeter of the central part of the northern mountain edge.

The final mountain in this group of 3 is Uranius Tholus lying to the north. It is much smaller than the other 2 and more circular in shape. I was able to locate a point (coordinates 262.4093° E 26.2088° N) on this mountain which was exactly $(R/4)(\sqrt{3}/\sqrt{5})$ = 657.67 km from the Fesenkov Crater and $2\pi R/(60\sqrt{2})$ = 251.48 km from the Paros Crater. The sacred distance from the Fesenkov Crater mimics the pattern of using ratios of sacred numbers such as the now familiar ratio e/π. The denominator of the sacred distance from the Paros Crater is very close to the sum of the denominators for the sacred distances to the other 2 mountains from the Fesenkov Crater (84.85 vs. 36+49=85). This surveyed centre for Uranius Tholus was found to be only about 1 km from the centre of a circle which followed the perimeter of the mountain, and therefore is very credible.

When I measured the distances between the mountains based on their survey centres, I found that the distances between Uranius Mons and the other 2 mountains (245.51 km to Uranius Tholus, 244.51 km to Ceraunius Tholus) differed by only 1 km, indicating that once again, we had an isosceles triangle. A difficulty arose when I searched for a sacred distance that would be meaningful. There were 3 very good contenders. The first was $(R/16)(\pi/e)$ = 245.32 km, the second, $R/(8\sqrt{3})$ = 245.10 km and the third, $(R/12)(e/\pi)$ = 244.88 km. I thought that perhaps the issue could be solved by finding a match with the distance between Uranius Tholus and Ceraunius Tholus. No match could be found for the second formula but unfortunately a match could be found for the other two: $(R/16)(\varphi/e)$ = 126.35 km and $(R/12)(1/\sqrt{5})$ = 126.57 km so that did not settle the issue. In the end, I chose the R/12 pair of numbers for my diagram since they gave the smallest sum of squared deviations (0.520 vs. 0.758 km²) from distances determined from the survey centres for the 3 mountains. The largest individual distance deviation for R/12 was only 0.63 km. However, the hypothetical architects may have also intended the R/16 pair. The final answer cannot be determined from present information. This elegant isosceles triangle is shown in Fig. 7.4. Besides the isosceles nature of the triangle and the matching of R/12 (or R/16) in sacred distances of the

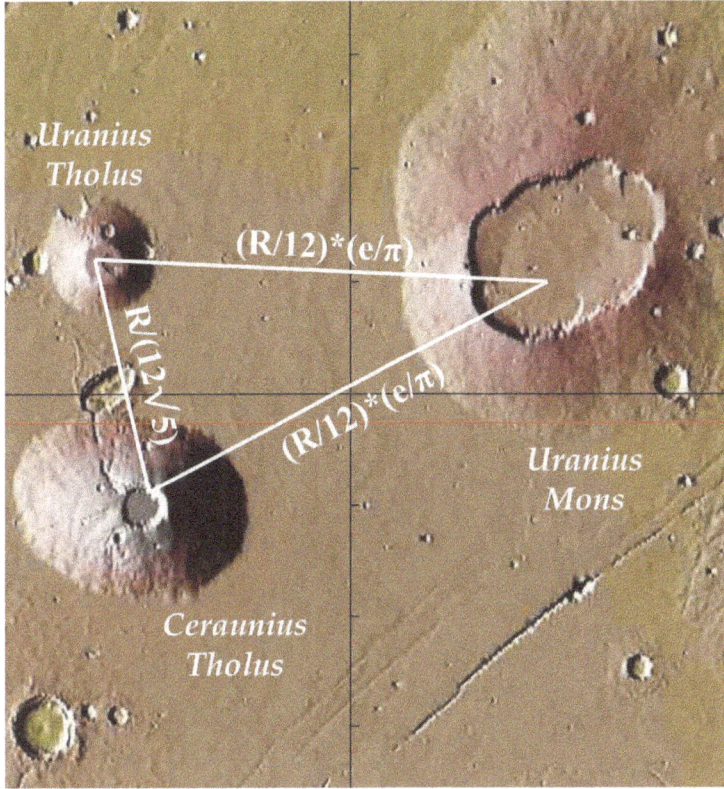

Fig. 7.4: *Sacred geometry triangle of the 3 mountains on the northern portion of the Tharsis Rise. Note the symmetry of the 3 sides in the use of the number 12. This is another example of the extensive use of e/π, √5 and the isosceles triangle in the arrangement of mountain positions. The side between Uranius Tholus and Uranius Mons has a bearing very close to -φ radians (92.707 degrees). USGS Astrogeology.*

sides, an interesting feature of the triangle is that the side between Uranius Tholus and Uranius Mons has a clockwise bearing angle (92.794°) that is only about 5 min of a degree larger than φ radians = 92.707°.

Tharsis Tholus

The next mountain that I surveyed from the Paros and Fesenkov Craters was Tharsis Tholus. This mountain measures about 155 km long by 125 km wide, and is about 9 km high[5] which is approximately the height of Mount Everest. The present version of Tharsis Tholus is actually 2 volcanoes rather than a single peak. The largest peak is located on the northwest end of the mountain group and a smaller peak occurs on the southeast end. In the rest of the book I will refer to these peaks as simply Tharsis Tholus North and Tharsis Tholus South. The outside wall of the west side of the northwest peak has collapsed and disappeared. The oldest part of Tharsis Tholus is very ancient, being estimated to be 3.82 billion years old[5].

A sacred geometry survey centre was first located for the northern peak of Tharsis Tholus (see Fig. 7.5). The coordinates for this point (268.7222° E

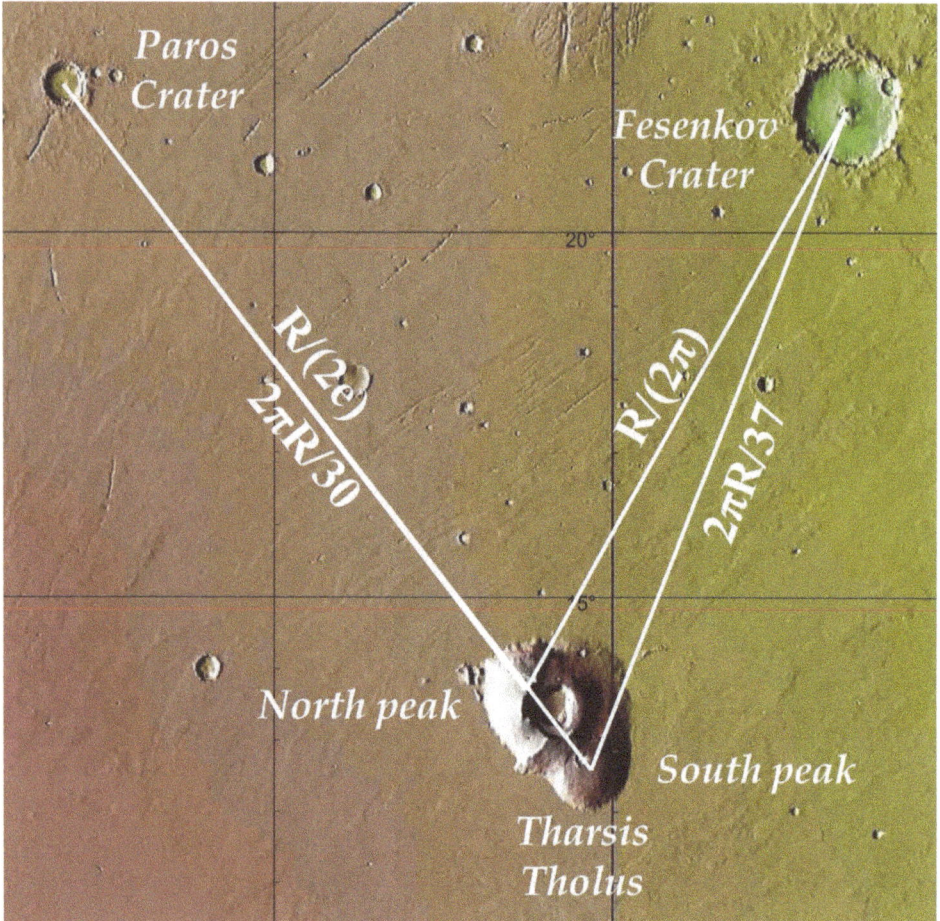

Fig. 7.5: *Determination of mountain centres for the 2 peaks of Tharsis Tholus using the Paros and Fesenkov craters as survey craters. The sacred distances for the 2 peaks are almost identical in bearing from the Paros Crater. Note the symmetry in sacred distances from the 2 craters for both the northern and southern peaks. USGS Astrogeology.*

13.7122° N) were found to be very close to the centre of a circle fitted to the northern perimeter of the north northwestern portion of the mountain. The sacred distance from this location to the centre of the Fesenkov Crater is the now familiar number $R/(2\pi)$ or 540.52 km. The sacred distance from the centre of the Paros Crater was found to be $R/(2e)$ or 624.69 km which is an elegant way to pair the distances from the 2 craters. Thus the northern Tharsis Tholus mountain peak is fitted very well.

The southern mountain peak of Tharsis Tholus could be fit to another matching pair of sacred numbers, being $2\pi R/37$ or 576.73 km distant from the Fesenkov Crater and $2\pi R/30$ or 711.30 km distant from the Paros Crater. The survey centre for this peak (coordinates 269.6923° E

12.5955° N) was very closely matched to the centre of a circle which followed the south southeast perimeter of the mountain so it had a high degree of credibility. Incidentally, the bearing to the Paros Crater in the counterclockwise direction from due north (38.438°) is only 14.6 min of a degree more than from the northern peak (38.195°), suggesting that the crater might have been positioned to be in alignment with the 2 mountain peaks. The reason for this alignment might have been to create the sacred bearing (counterclockwise) of $asin(1/\varphi) = 38.173°$ which is only 1.3 min less than that of the northern peak and 15.9 min less than that of the southern peak. The southern peak is located at a distance of 86.70 km from the northern peak.

When the distances to other mountains were calculated from the 2 newly surveyed locations of Tharsis Tholus, a number of surprises awaited me. To begin with, the southern peak was found to be related to Ascraeus Mons and Arsia Mons by a pair of symmetrical sacred numbers. This created the sacred triangle shown in Fig. 7.6. The distance to the survey centre for Ascraeus Mons was only 0.87 km more than $2\pi R/(16\varphi)$ or 824.26 km, and that to the survey centre for Arsia Mons was 2.64 km more than $2\pi R/(6\varphi)$ or 2198.03 km. The third side of the triangle would be the length between Ascraeus Mons and Arsia Mons which I previously established to be 0.05 km more than $R/\sqrt{5}$ or 1518.82 km. Remarkably, this latter side turns out to be almost exactly one-half of the sum (i.e., the average) of the other 2 sides (100.5% of one-half the sum of the theoretical distances, 100.4% if the actual measured distances were used).

The northern peak is hardly to be outdone by the southern peak in generating interesting sacred distances to other mountains. I will focus for the present on the sacred distances to Ulysses Tholus ($\pi R/(4\sqrt{2})$ =1886.11 km, deviation from theoretical = -0.76 km) and to Ceraunius Tholus ($e^2 R/36$ = 697.07 km, deviation = -0.31 km). The reason that these 2 mountains are interesting is that another isosceles triangle is formed when we consider that the sacred distance between Ulysses Tholus and Ceraunius Tholus is also $\pi R/(4\sqrt{2})$ km with a deviation of −1.38 km). This triangle is depicted in Fig. 7.7. Although slightly narrower than one-half the width of the bisected isosceles triangle of Chapter 5 joining the Tharsis Montes with Olympus Mons, this isosceles triangle is actually taller.

There are many more sacred triangles created with the northern peak of Tharsis Tholus. Several of these will be shown in the next chapter with the analysis of other mountains to the north, west and southwest of Ascraeus Mons.

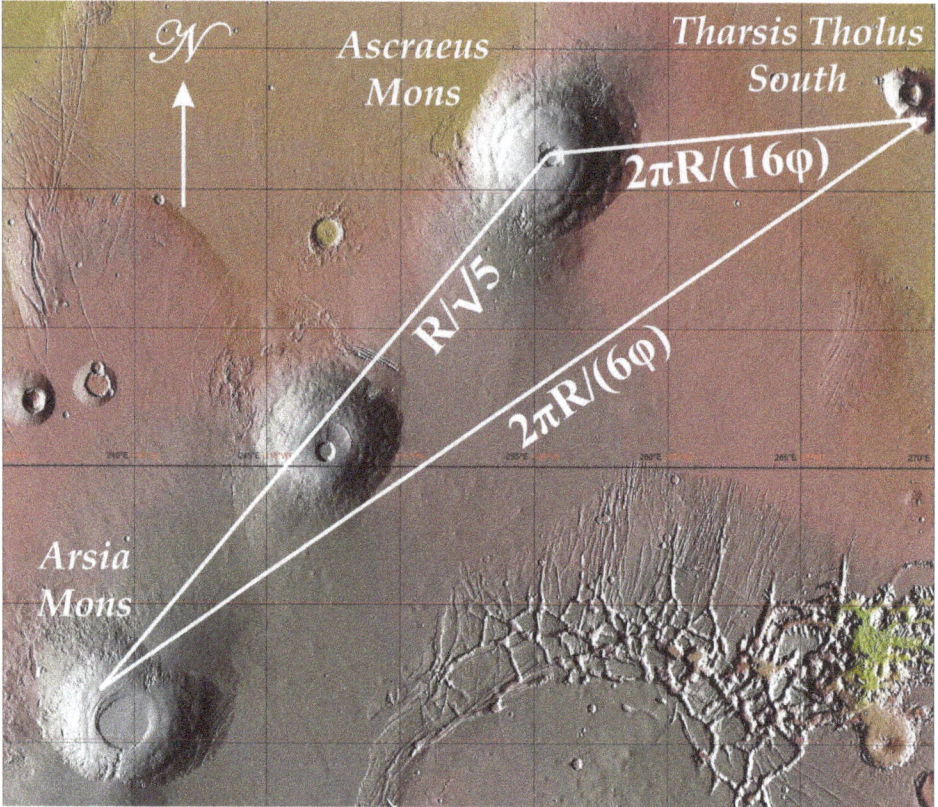

Fig. 7.6: *Sacred geometry triangle involving the southern peak of Tharsis Tholus. This triangle has the property of the side between Ascraeus Mons and Arsia Mons being almost equal to 1/2 the sum of the other 2 sides. Note the symmetry in the use of the equatorial circumference of Mars and φ in the 2 sides emanating from the Tharsis Tholus South peak. USGS Astrogeology.*

References

1. *List of mountains on Mars by height.*
    http://en.wikipedia.org/wiki/List_of_mountains_on_Mars_by_height

2. *The Biblis Patera Volcano. Mars Express. ESA. http://www.esa.int /Our_Activities/Space_Science/Mars_Express/The_Biblis_Patera_volcano*

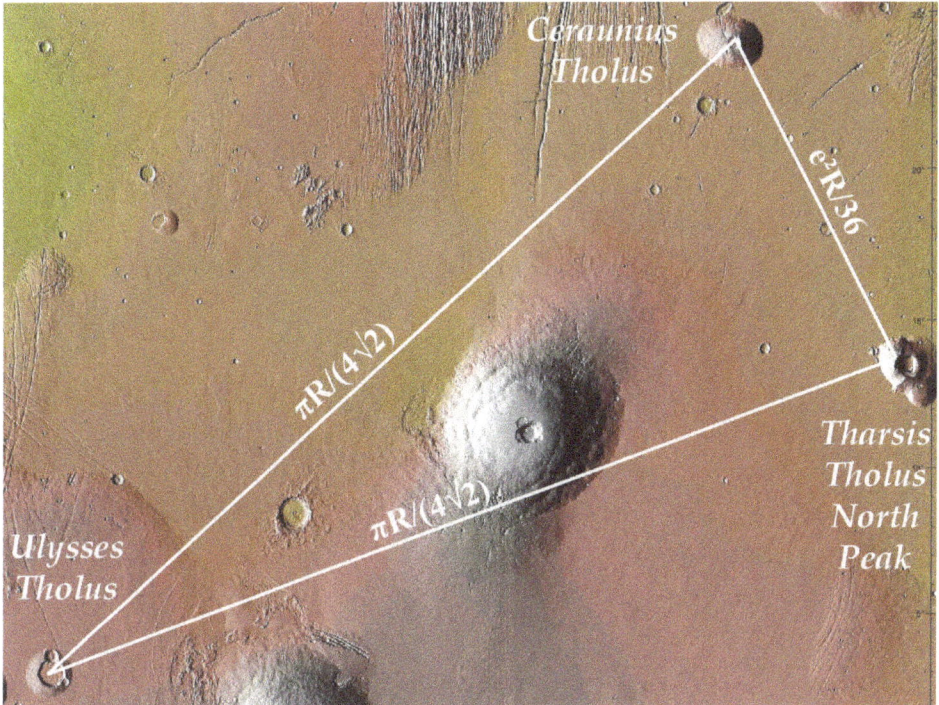

Fig. 7.7: *Sacred isosceles triangle emanating from Ulysses Tholus to Ceraunius Tholus and the northern peak of Tharsis Tholus. This triangle is taller than the one joining the Tharsis Montes to Olympus Mons, but is less than one-half its width. The mountain inside the triangle is Ascraeus Mons. USGS Astrogeology.*

3. *Characterization of Major Volcanic Edifices on Mars Using Mars Orbiter Laser Altimeter Data. J. W. Head, N. Seibert, S. Pratt, D. Smith, M. Zuber, S.C. Solomon, P.J. Morgan, J.B. Garvin. http://www.researchgate.net /publication/252260845_Characterization_of_Mars_Using_Mars_Orbiter _laser_Altimeter_Data*

4. *http://volcano.oregonstate.edu/volcanic-cones-mars*

5. *Cycles of edifice growth and destruction at Tharsis Tholus, Mars. T. Platz, P.C. McGuire, S. Münn, B. Cailleau, A. Dumke, G. Neukum, and J.N. Procter. http://www.researchgate.net/publication/233426405_Cycles_of_edifice_ growth_and_destruction_at_Tharsis_Tholus_Mars*

8

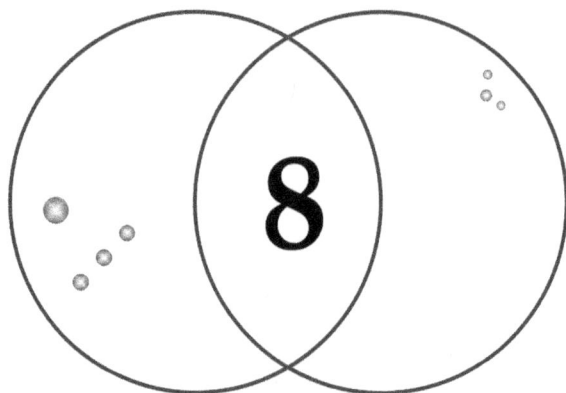

Lesser Mountains Part II. North, South and Centre

We have touched on the smaller mountains that seem to be arranged in groups. There are still several other 'small' mountains that tend not to receive as much attention as the giants and which exist apart in their own separate areas. This chapter will analyse these mountains in relation to themselves and to the other mountains we have already covered. I have found that they comprise a very important segment of the mountain masterpiece that is laid out across the planetary landscape. To start with I will look at 2 northern mountains, proceed to a central mountain, and then end with Apollinaris Mons to the southwest as well as 2 large craters to the west of Olympus Mons.

Issedon Tholus

If we wander further north of Uranius Tholus by about 10 degrees, we come across a very small mountain called Issedon Tholus. I only noticed it when I started looking for survey craters for Alba Mons. Very little information is available on this mountain, and everything that I know about it mostly comes from my own examination of the USGS map. Its elevation above datum is listed at 826 meters[1]. The makers of the USGS map used a tiny filled white square to mark out the location of the peak for this mountain. They placed the square on top of what appears to be a small caldera whose centre was measured to have the coordinates of 265.1574° E 36.0006° N. I determined the coordinates based on a higher definition THEMIS (Thermal Emission Imaging System aboard the Mars Odyssey spacecraft) image of Issedon Tholus which (at high magnification) shows the 36.00° grid line passing virtually right through the centre of this peak/caldera (http://planetarynames.wr.usgs.gov/Feature/2745).

Most amazing though was the discovery that the perimeter of the base of Issedon Tholus is rectangular rather than rounded as we would expect for a mountain. At first I thought that the perimeter was square and proceeded to fit it according to this assumption (see dashed square in Fig. 8.1). The sides of the square had bearing angles of 45° in the clockwise or counterclockwise directions so that its vertices pointed to the 4 cardinal directions of north, east, south and west. When I later examined the higher definition THEMIS image it became obvious that the perimeter extended quite a bit further to the southwest and actually had the shape of a golden rectangle (length is 1.6180 times longer than the width) rather than a square (Fig. 8.2). I will delay the discussion of my fitting procedure for the rectangle and its square portion until the second book of the *Intelligent Mars* series where I give a much more detailed presentation of the Issedon Tholus structure. The delay is necessary since many more discoveries and concepts are needed to be covered before a fuller explanation can be undertaken.

The rectangular nature of the Issedon Tholus perimeter of course leads one to suspect that we are dealing with an artificial structure, something more like a pyramid than a mountain. The base of this pyramidal structure appears to extend beyond the southwest boundary of the fitted square, but it is not clear if it goes all the way to the southwest side of the golden rectangle. If we assume that the base does indeed extend to the full dimensions of the rectangle, the pyramid would have a width of about 52 km and a length of about 85 km. The width alone is a staggering 225 times the side length of the great pyramid of Giza in Egypt. Assuming its elevation of 826 meters to be an approximation of its height, it would be about 5.6 times the height of the great pyramid. In terms of volume, it is approximately 470,000 times the size of the Giza pyramid! Nothing seems to come small on Mars and I am starting to run out of the superlatives needed to describe it.

A survey centre for Issedon Tholus could be obtained by finding the intersection point of the sacred distances of eR/(6φ) = 950.93 km from the Fesenkov Crater and R/4 = 849.05 km from the Paros Crater. The surveyed location (265.2990° E 36.0037° N) has some very interesting

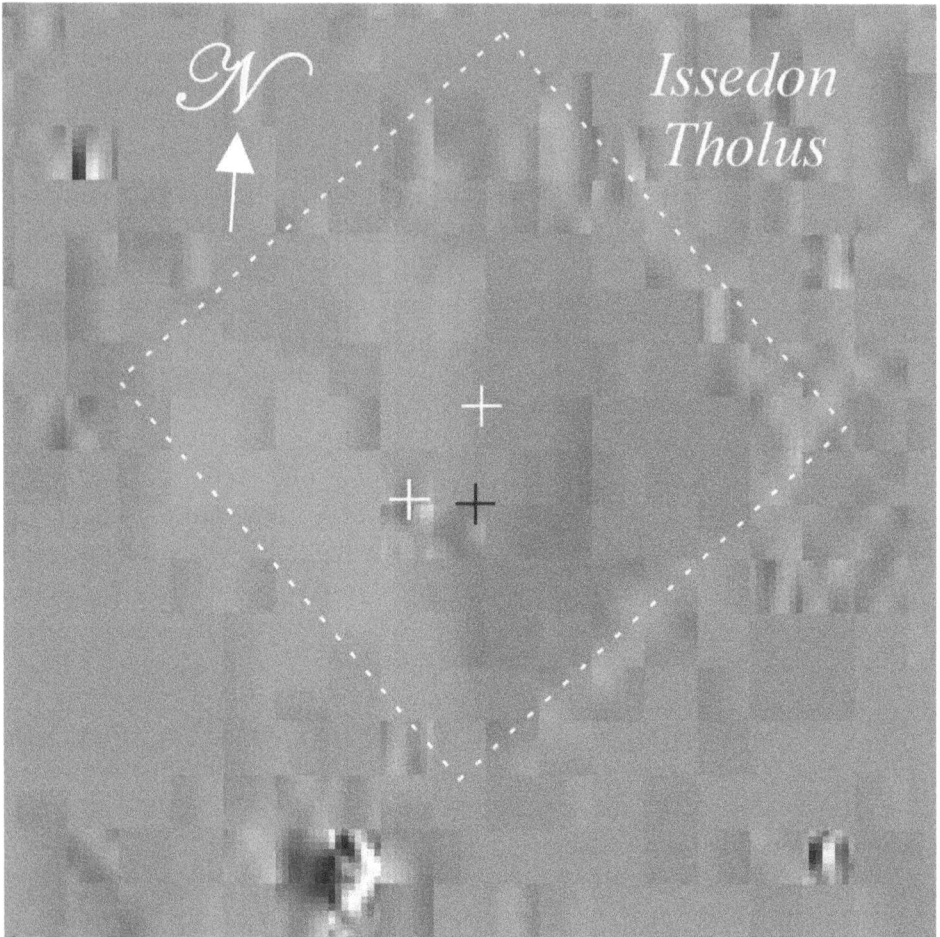

Fig. 8.1: *Issedon Tholus with its base perimeter fitted to a dashed square. Its peak (caldera) is marked with a yellow cross (left cross) which replaces the small filled white square placed by the makers of the USGS map. The survey centre is marked by a black cross (lower right cross) and the centre of the square is marked by a white cross (highest cross). Each side of the square is approximately 52 km in length. The NE and SE sides are reasonably visible whereas the NW side is much more difficult to make out. The SW side had to be drawn based on information from the other 3 sides since the map does not have the data necessary to define it, and may have a virtual presence only. USGS Astrogeology.*

properties. To begin with, its latitude is only 13.3 seconds of a degree more than 36.0000° so for all intents and purposes it is at the same latitude as the Issedon Tholus peak/caldera and lies 6.79 km away to the east. The longitude of the survey centre is 8.5 minutes of a degree to the east of the Issedon Tholus peak/caldera, and appears to have the same longitude as the north and south vertices of the dashed square in Fig. 8.1. The orientation of Issedon Tholus is analogous to the great pyramid of

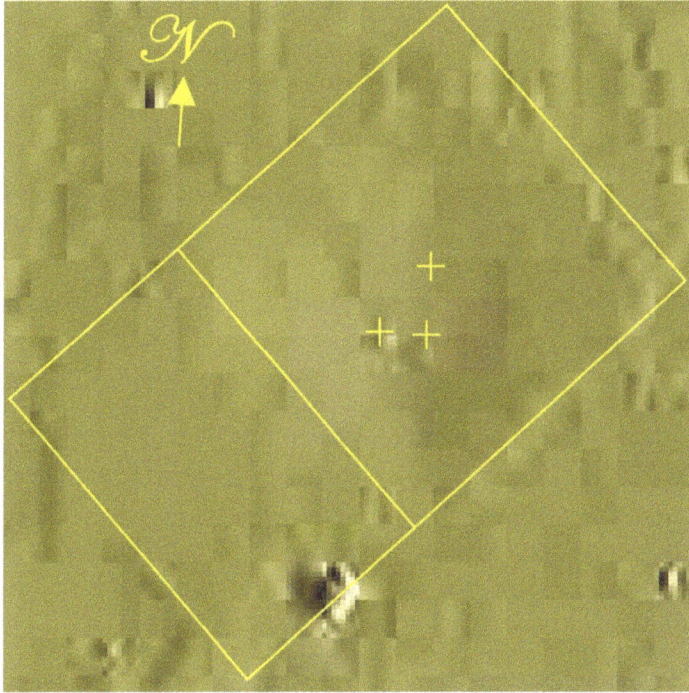

Fig. 8.2: *The square in Fig. 8.1 extended to a golden rectangle. The southwest side of the rectangle fits the bottom edge of a long cliff and the southeast side runs along the bottom edge of the wall of a small crater. The portion of the rectangle extending beyond the square is itself a golden rectangle. Issedon Tholus may actually be a pyramid whose base edges correspond to the course of the large golden rectangle. USGS Astrogeology.*

Giza whose northern side faces due north to a precision of 3 minutes of a degree, except with Issedon Tholus it is the northern vertex that points north in combination with the southern vertex. The centre of the square perimeter also has the same longitude as the Issedon Tholus survey centre. The matching of the longitudes of the north and south vertices and centre of the square with the longitude of the survey centre gives credibility to the square portion of the golden rectangle even though it may be virtual rather than structural. Another interesting feature of the survey centre for Issedon Tholus is that it is about 9.5 km south of the centre of the square. Since it has the same latitude as the peak (caldera) of Issedon Tholus, both the survey centre and the peak are off-centre for the square. They also do not sit at the middle of the golden rectangle. Hence, Issedon Tholus is unlikely to be a symmetrical pyramid like the Giza pyramid but rather an asymmetrical pyramid.

The survey centre for Issedon Tholus thus has the properties of being at the same latitude as the caldera and sharing the same longitude as the centre of the square. When I measured the distance of this site to other mountains and craters, I found that the Issedon Tholus 'mountain' was definitely part of the matrix of sacred geometry on Mars. It has a strong association with the triple mountain group described in the previous chapter, as well as with the northern peak of Tharsis Tholus. It is also associated by a sacred distance with the survey coordinates of Alba Mons.

Now let's look at the actual sacred distances to the other mountains. Alba Mons (see below) is only 0.80 km further than the very familiar sacred distance of $R/(2\sqrt{5})$ or 759.41 km from the survey centre for Issedon Tholus. Uranius Tholus is only 1.62 km closer, and Uranius Mons, 1.45 km closer than the sacred distance of $R/(4\sqrt{2})$ or 600.37 km. The third mountain of this group, Ceraunius Tholus, is only 0.62 km further than the sacred distance of $eR/(8\varphi)$ or 713.20 km. Finally, the northern peak of Tharsis Tholus is only 0.18 km further than the sacred distance $2\pi R/16$ or 1333.68 km.

In Fig. 8.3 there are 2 triangles involving Issedon Tholus. The first of these is an isosceles triangle with Uranius Mons and Uranius Tholus. The 2 equal sides are set to sacred distances of $R/(4\sqrt{2})$ km as mentioned above. Alternative sacred distance formulae (where R' is the northern polar radius) are $\sqrt{2}\pi R'/25 = 600.00$ km, $\pi R/(11\varphi) = 599.46$ km and $e\varphi R/25 = 597.49$ km which are closer to the actual values but being more complex formulae, I decided to use the $R/(4\sqrt{2})$ km formula in Fig. 8.3. The base of the triangle is close to $(R/12)(e/\pi)$ km in length. At first all we see in this triangle are the sacred numbers and its isosceles nature. But if we probe deeper, we discover that the average length of the 2 equal sides is only 0.3% less than $1/100^{th}$ of the square of the base! Or put another way, the base is ten times the square root of one of the equal sides. The architects may have intended this relationship to fit in with the fact that 10 times the latitude of Issedon Tholus is 360 degrees, the total number of degrees around the planet.

The second triangle in Fig. 8.3 is very unusual in that it is extremely flat, being only about 5.5 km in height at the apex occurring on Uranius Mons. The length (1333.86 km) of the base of the triangle between Issedon Tholus and the northern peak of Tharsis Tholus has 2 possible sacred formulae. The closest match is $2\pi R/16 = 1333.68$ km and the next closest is $5R/(9\sqrt{2}) = 1334.15$ km. The side (734.75 km) between Uranius Mons and the northern peak of Tharsis Tholus is extremely close to the sacred formula $(R/4)(e/\pi) = 734.64$ km shown in Fig. 8.3. Interestingly, this sacred distance is exactly 3 times the sacred distance between Uranius Mons and Uranius Tholus.

Each of the mountains in the triple group of Ceraunius Tholus, Uranius Tholus, and Uranius Mons is now shown to make a triangle side whose sacred distance has $\sqrt{2}$ in the denominator (remember the isosceles triangle from Ulysses Tholus to Ceraunius Tholus and Tharsis Tholus North in Fig. 7.7 which used $\pi R/(4\sqrt{2})$ km). Thus these 3 mountains seem to celebrate $\sqrt{2}$ in the way that Biblis Tholus and Ulysses Tholus celebrate $\sqrt{3}$, and the Tharsis Montes and Olympus Mons celebrate $\sqrt{5}$. These square root values are extremely important irrational numbers in sacred geometry. Another remarkable thing to note about all of these triangles, including the isosceles triangle from Ulysses Tholus, is that the denominator for every side is divisible by the number 4. Since $\sqrt{2}$ is a factor in the diagonal of a square

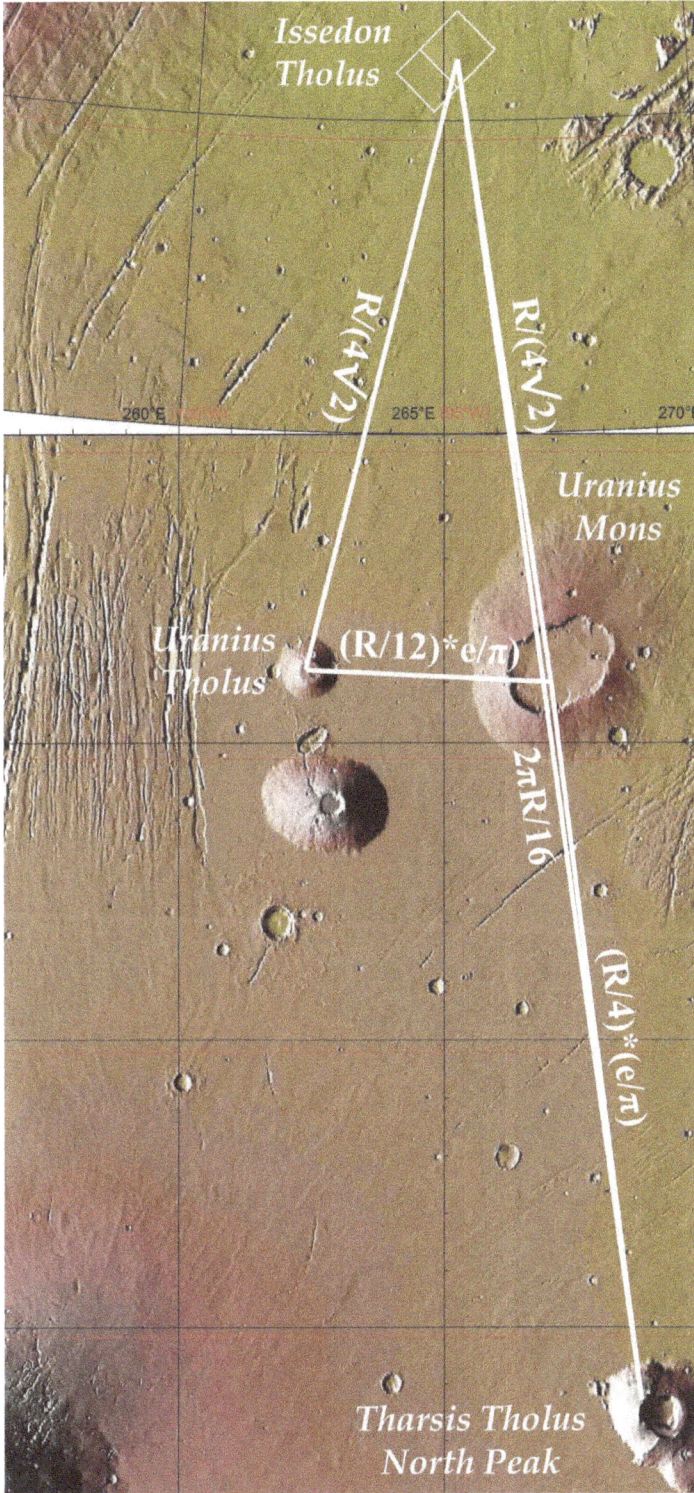

Fig. 8.3: *Sacred geometry triangles for Issedon Tholus (white golden rectangle marks its theoretical perimeter). The triangle from Issedon Tholus to the northern peak of Tharsis Tholus is extremely flat, being only about 5.5 km at its apex at Uranius Mons. The triangle emanating from Issedon Tholus to Uranius Mons and Uranius Tholus is isosceles. Note also that the distance from Uranius Mons to Uranius Tholus is one-third the distance from Uranius Mons to the northern peak of Tharsis Tholus. USGS Astrogeology.*

and 4 is the number of equal sides in a square, the sacred distance formulae associated with these 3 mountains may allude to the square portion of Issedon Tholus. The alternative formulae of $\pi\sqrt{2}R'/25$, $\pi R/(11\varphi)$ and $e\varphi R/25$ km for the equal sides of the isosceles triangle from Issedon Tholus to Uranius Tholus and to Uranius Mons have strong associations with the pentagram (see Chapter 9) and hence, with the golden rectangle of Issedon Tholus. All the formulae were likely intended, each conveying an appropriate meaning.

Alba Mons

I now come to what has been termed the most massive mountain in the solar system in terms of area and volume. Alba Mons is situated more than 28 degrees north of Ascraeus Mons at a longitude about 6 degrees further west. It gets its volume not from its height (about 7 km above datum) but from its enormous diameter (a whopping 2700 km by some accounts!). The southern slope of Alba Mons seems to have been lifted by the up-warping of the Tharsis Rise, since the dome of this mountain is tilted. Its location is antipodal to the Hellas basin suggesting a link between these 2 gigantic features. The total volume of effusive deposits from Alba Mons has been recently calculated by Ivanov and Head[2] to be about 2.44 to 2.64 million cu km which might actually make it slightly smaller in volume than Olympus Mons. Based on an average profile of Olympus Mons by MacGovern and Morgan[3], I made a rough estimate of about 2.7 million cu km for the volume of Olympus Mons.

After not being able to locate any satisfactory survey craters for Alba Mons, I decided to follow a hunch and used the survey centres of Olympus Mons and Ascraeus Mons to locate the intersection point of the sacred distance $R/2 = 1698.10$ km from both mountains. The resulting surveyed centre (249.6749° E 39.4492° N) lay just inside the southern edge of the caldera complex about 18 kilometers west of the deep southeast caldera (see Figs. 8.4 & 8.6), and had many features which gave it a high degree of credibility for the original centre of Alba Mons. As mentioned above, it was only 0.80 km further than the sacred distance of $R/(2\sqrt{5})$ or 759.41 km from the survey centre for Issedon Tholus. It was only 0.54 km farther than $\pi R/(3e) = 1308.36$ km from the AscSC1 Crater and 0.01 km more than $3R/8 = 1273.57$ km from the AscSC2 Crater. It was only 0.18 km closer than $\sqrt{5}R/4\varphi = 1173.35$ km from Uranius Mons, and 1.18 km further than $8R/27 = 1006.28$ km from Uranius Tholus. And if this wasn't enough, it was found to be so close to the sacred distance of $\pi R/(3e) = 1308.36$ km from the eastern centre of Jovis Tholus that I used it to survey this centre (see below). This is the same sacred distance as the AscSC1 Crater from Alba

Mons! With these credentials, I was confident that at long last I found a fit to my original equilateral triangle model. This surveyed centre of Alba Mons creates an enormous "equilateral" triangle although the side between Olympus Mons and Ascraeus Mons is about 30 km less than the theoretical distance of R/2 km due to the adjustment required to fit the sacred bisected isosceles model as discussed in Chapter 5. Strictly speaking, this triangle is actually isosceles in nature, and can be seen in Fig. 8.6.

Using these survey coordinates of Alba Mons, I could now construct sacred triangles for this mountain. What I found was an unusually high number of sacred distances to other sites which were simple functions of the equatorial radius R. We have already seen R/2 km distances between Alba Mons and either Olympus Mons or Ascraeus Mons. Now in Fig. 8.4, we can see 2 more triangles with sides equal to R/3, R/4 and R/(2√5) km emanating from Alba Mons, all within 1 km of their theoretical distances. The distance of R/3 = 1132.06 km is between Alba Mons and the centre of the Ceraunius Tholus Caldera (262.9032° E 23.9674° N). The distance of R/4 = 849.05 km is between Alba Mons and the Ayacucho Crater (267.9781° E 38.1812° N) which is a tiny crater about 2.5 km in diameter situated about 180 km to the northeast of Issedon Tholus. Despite its size, this crater seems to be an important site on Mars and it is discussed in greater length in the 2nd book of this series. It appears to be positioned on top of a structure similar to Issedon Tholus and may actually be a marker of a pyramid peak rather than a true crater. The distance between Issedon Tholus and the Ceraunius Tholus Caldera is 0.61 km less than 16R/75 = 724.52 km, and the distance between the Ayacucho Crater and the Ceraunius Tholus Caldera is 0.70 km more than 2πR/(14√3) = 880.00 km.

It should be noted that the upper portion of the map in Fig. 8.4 is curved, resulting in the northward displacement of Alba Mons by about 1 degree. In a Mercator projection map, the position of Alba Mons would be lower and hence give a more accurate picture. This figure is intended only to orient the reader.

You may have noticed that I have used the caldera centre rather than the survey centre for distance measurements to Ceraunius Tholus. I have discovered that in addition to survey centres, the caldera centres of mountains have been used as important sites to create very meaningful sacred distances or coordinate values. For this reason, I have determined the coordinates of the centres of calderas of mountains where feasible, i.e., where calderas offer a distinct circular shape. On Olympus Mons, I measured 2 calderas for which the centres could be determined (Fig. 8.12). For the remainder of the book, mountain calderas will be considered in addition to mountain survey centres in the determination of significant sacred distances to other points of interest.

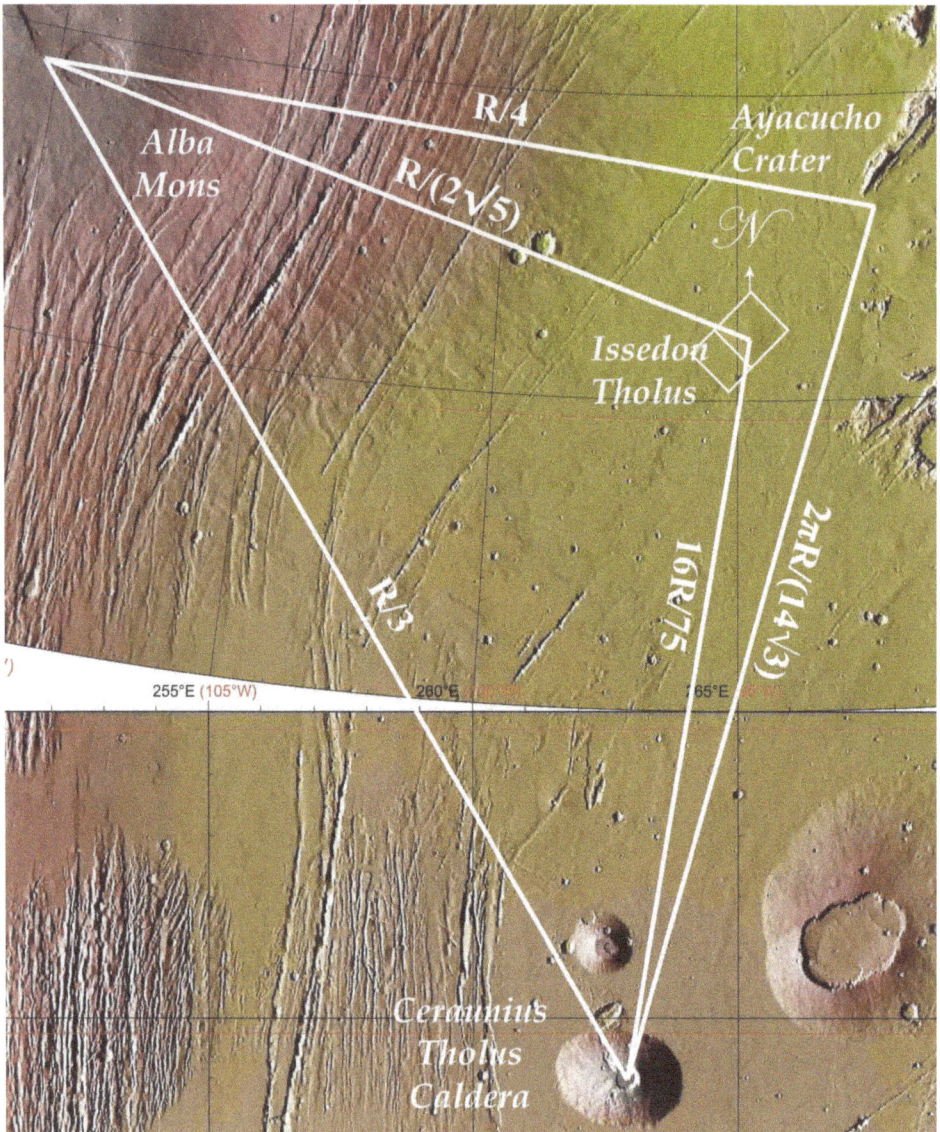

Fig. 8.4: *Two sacred triangles for Alba Mons. The Ayacucho Crater is too small to be visible at this magnification. The triangle sides emanating from Alba Mons are simple functions of the planetary radius. Alba Mons is displaced northerly by about 1° due to upper map curvature. USGS Astrogeology.*

Jovis Tholus

Jovis Tholus is another of those small Martian mountains that seems to be rather insignificant. It is only about 2 km in height and about 80 km long (east-west) by 60 km wide (north-south). It is estimated to be 2.3 to 3.5 billion years old[4] and has a caldera complex about 40 km in diameter on the west side. It actually has 2 centres, one based on its eastern perimeter and the other based on its western perimeter (Fig. 8.5). When I tried to fit these 2 centres to the surveyed centres of other mountains and to crater centres, I discovered, as with Issedon Tholus, that this seemingly insignificant mountain was an important piece of the sacred geometry puzzle on Mars. Firstly, I found that I could survey the western centre of Jovis Tholus by determining the location equidistant from Olympus Mons and Ascraeus Mons by a value of $2\pi R/25$ or 853.56 km. This location (242.4932° E 18.2598° N) corresponded nearly exactly with the centre of a circle fitted to the western perimeter of the mountain. The eastern centre of the mountain could be surveyed by using the survey centre of Alba Mons and the AscSC1 Crater as survey reference points. This was accomplished by using the sacred distances of $(R/3)(e/\pi)$ or 979.52 km from the AscSC1 Crater and $(R/3)(\pi/e)$ or 1308.36 km from Alba Mons (see Fig. 8.9 below). Note that these 2 formulae are matched, with the last term of one equation being the inverse of the last term in the other. The calculated location had the coordinates 242.6544° E 18.2340° N and was very close to the centre of a circle fitted to the eastern perimeter of Jovis Tholus. The eastern and western centres are 9.20 km apart, and I will henceforth refer to them as Jovis Tholus West and Jovis Tholus East.

Some very interesting sacred triangles are created with Jovis Tholus West. In Fig. 8.6 we can see the huge "equilateral" triangle between Alba Mons, Olympus Mons and Ascraeus Mons mentioned previously. Within this "equilateral" triangle, 3 other triangles are formed with Jovis Tholus West. Two of the triangles are mirror images of each other and the third is an isosceles triangle joining Olympus Mons and Ascraeus Mons with 2 sides exactly equal to the survey value of $2\pi R/25$ km. All sides emanating from Jovis Tholus West incorporate the equatorial circumference of Mars,

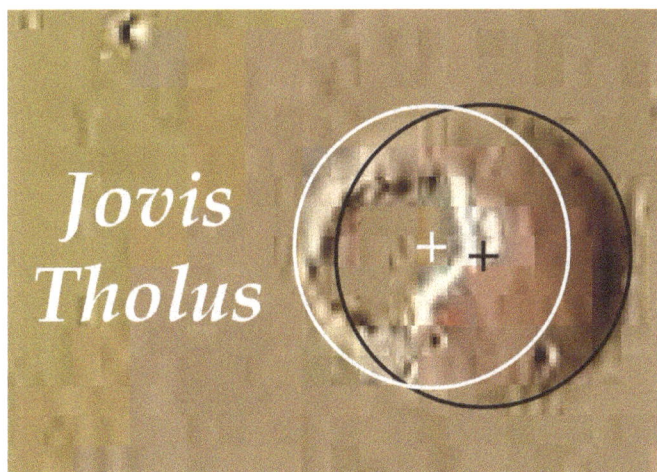

Fig. 8.5: *Jovis Tholus has a western survey centre (white cross) and an eastern survey centre (black cross). Circles from each centre align with the tholus western and eastern perimeters respectively. The 2 centres are 9.20 km apart. USGS Astrogeology.*

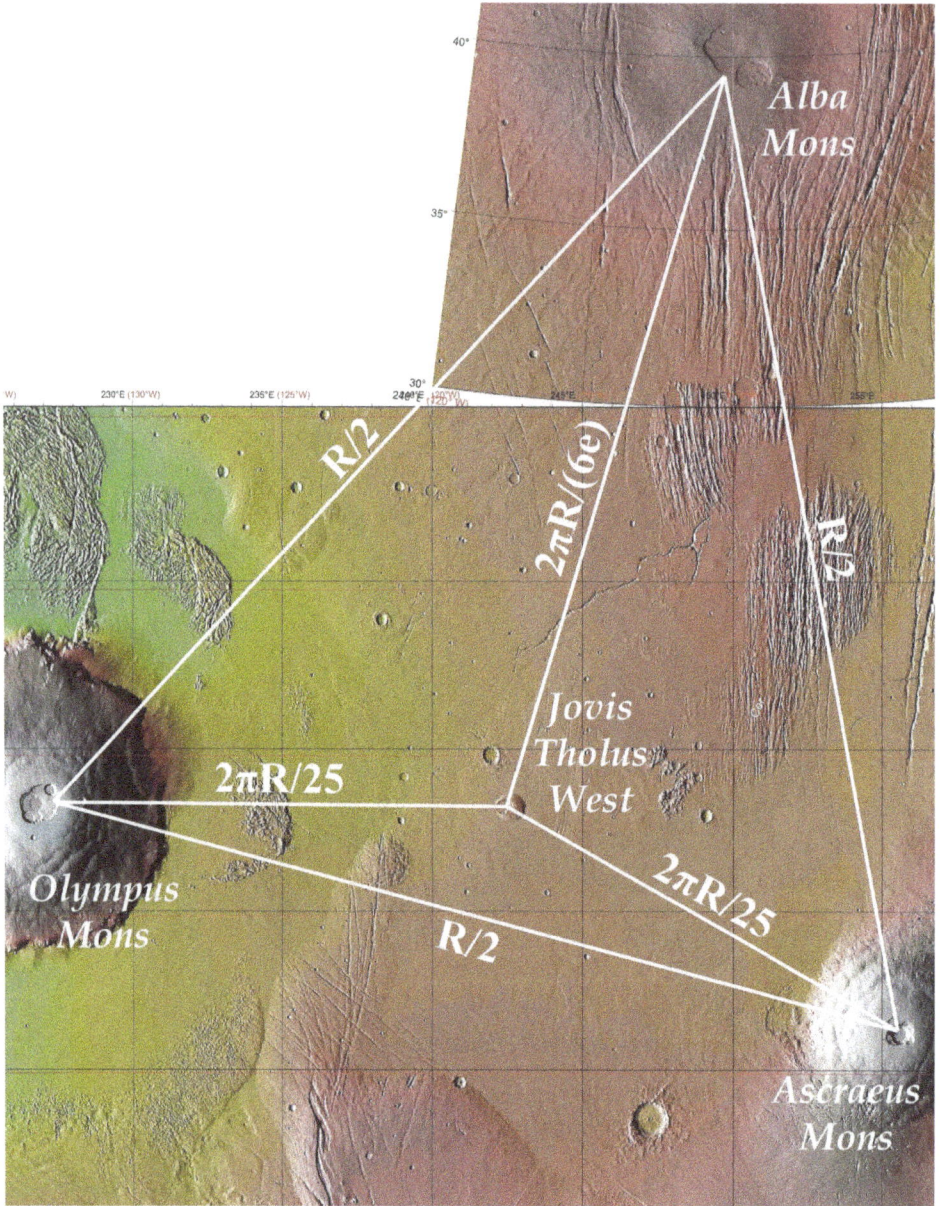

Fig. 8.6: *Sacred geometry triangles for the western centre of Jovis Tholus and for Alba Mons. The major triangle is "equilateral" with equal sides of R/2 km between Olympus Mons, Alba Mons and Ascraeus Mons. The side between Olympus Mons and Ascraeus Mons is actually about 30 km less than R/2 km to compensate for spherical distortion in the bisected isosceles triangle fitting the giant mountains (see Chapter 5). Enclosed in this huge triangle are 3 smaller sacred triangles involving Jovis Tholus West. Note the symmetry in the use of the equatorial circumference for sides emanating from Jovis Tholus West. USGS Astrogeology.*

including the side to Alba Mons which is 0.83 km more than the distance $2\pi R/(6e) = 1308.36$ km.

A host of triangles are created emanating from Jovis Tholus West to Uranius Mons, Uranius Tholus, Ceraunius Tholus, Ascraeus Mons and the northern peak of Tharsis Tholus (Fig. 8.7). To begin with, a narrow triangle created from Jovis Tholus West, Uranius Tholus and Ceraunius Tholus is almost isosceles in nature. The side between Jovis Tholus and Uranius Tholus is 3.43 km longer than $2\pi R/18$ or 1185.49 km and the side between Jovis Tholus and Ceraunius Tholus is 2.69 km shorter than $2\pi R/18$ km. The length of the triangle's base $(R/(12\sqrt{5})$ km) is 1/12th the distance between Jovis Tholus West and Tharsis Tholus North $(R/(\sqrt{5})$ km). The importance of the number 12 will be discussed in Chapter 12. Intersecting this "isosceles" triangle is a very narrow obtuse triangle extending from Jovis Tholus West to Uranius Mons with a base equal to $\sqrt{3}\pi R/13$ or 1421.54 km (deviation from theoretical = -0.20 km) and having its apex at Ceraunius Tholus. As seen in Fig. 7.4, the distance from Ceraunius Tholus to Uranius Mons is $(R/12)(e/\pi)$ km, and this formula also contains the number 12.

The third triangle emanating from the western centre of Jovis Tholus goes to Ceraunius Tholus and the northern peak of Tharsis Tholus. What is remarkable about this triangle is that the distance from Jovis Tholus West to Tharsis Tholus North is only 0.56 km short of the length $R/\sqrt{5}$ or 1518.82 km, the exact same distance for the base and height of the sacred geometry triangle in Chapter 5. The distance between Ceraunius Tholus and Tharsis Tholus North is only 0.31 km shorter than the sacred distance of $e^2 R/36$ or 697.07 km.

The fourth triangle from Jovis Tholus West goes to Uranius Mons and Tharsis Tholus North. This triangle is notable for the fact that the theoretical distance of $(R/4)*(e/\pi) = 734.64$ km between Uranius Mons and Tharsis Tholus North is virtually equal to one-quarter of the sum of the theoretical distances of the other 2 sides $[(1421.54 + 1518.82)/4 = 735.09$ km]. The figures using the actual distances show an almost exact equality [734.91 vs. $(1421.34 + 1518.26)/4 = 734.90$ km].

This brings us to the fifth Jovis Tholus West triangle, the one involving Ascraeus Mons and Uranius Mons. The distance between Ascraeus Mons and Uranius Mons is 3.23 km greater than the sacred distance R/π or 1081.04 km which is twice the sacred distance of $R/(2\pi)$ km, a very popular distance for the survey craters for the Tharsis Montes. The other 2 sides have been discussed above.

The last triangle in this group does not involve Jovis Tholus. It emanates from Tharsis Tholus North and joins with Uranius Mons and Ceraunius Tholus. The theoretical distance of the short side between Uranius Mons and Ceraunius Tholus $[eR/(12\pi)$ km] is equal to one-third

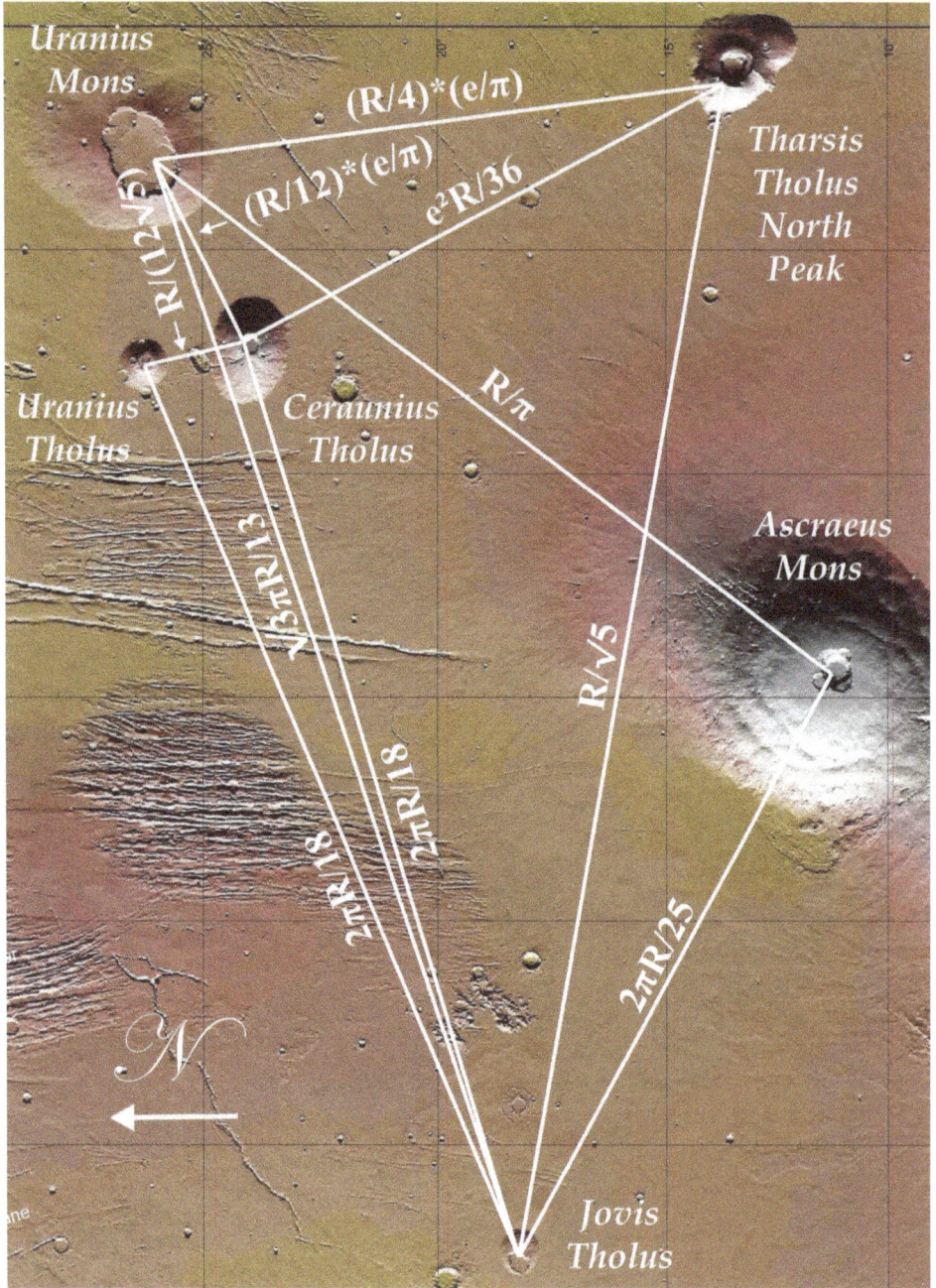

Fig. 8.7: *Five sacred geometry triangles emanating from the western centre of Jovis Tholus. A sixth emanates from the northern peak of Tharsis Tholus to Uranius Mons and Ceraunius Tholus. The triangle between Jovis Tholus West, Uranius Mons and Tharsis Tholus North has the special property of the short side being almost exactly equal to one-quarter of the sum of the 2 long sides. Note that north points to the left side of the page. USGS Astrogeology.*

of the distance between Tharsis Tholus North and Uranius Mons [$eR/(4\pi)$ km]. The distance of $e^2R/36$ km mentioned above between Ceraunius Tholus and Tharsis Tholus North completes a very interesting series of denominators in the 4 sacred lengths emanating from Ceraunius Tholus. They all contain integers (i.e., 12, 18, and 36) which are factors of 36 and 72. These latter two values are the number of degrees in the angles of a star point in a pentagram, a geometric figure in which the golden mean φ plays a prominent role (see Chapter 9).

While Fig. 8.7 may be somewhat challenging to take in, it provides an insight into how integrated all of the mountains and some craters really are. There is another interesting triangle emanating from the western centre of Jovis Tholus shown in Fig. 8.8. This one is composed of Pavonis Mons and an unnamed crater which I termed the AscSC1a Crater. The latter is a crater which could be used as a survey crater for Ascraeus

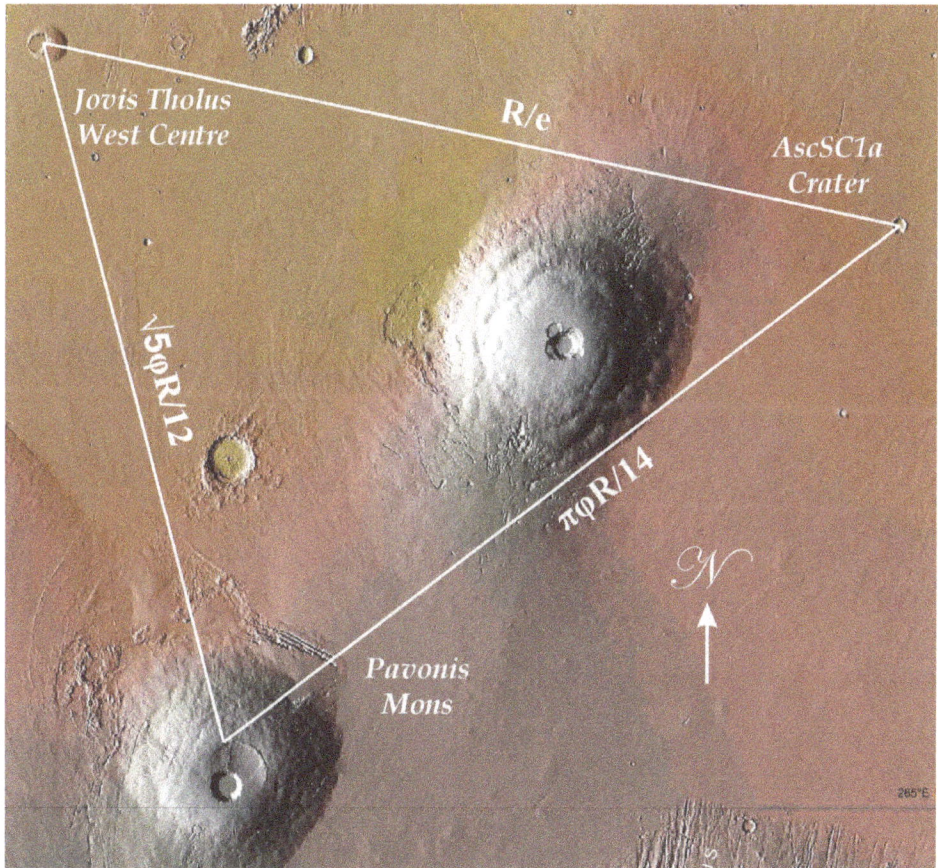

Fig. 8.8: *Sacred geometry triangle from Jovis Tholus West. The distance formula from Jovis Tholus West to the AscSC1a Crater is simply R/e km. The distance formula from Jovis Tholus West to Pavonis Mons emphasizes the golden mean. USGS Astrogeology.*

Mons (see Chapter 9) but it is at a different distance from Ascraeus Mons than the $R/(2\pi)$ km of the other survey craters. I discovered it much later than AscSC1 and AscSC2 so I called it AscSC1a in order to keep the original names for the other survey craters for Ascraeus Mons. The distance from Jovis Tholus West to AscSC1a is only 0.83 km greater than $R/e = 1249.39$ km. The distance from the survey centre of Pavonis Mons to the AscSC1a Crater is only 0.70 km more than $\pi\varphi R/14 = 1233.11$ km, and to Jovis Tholus West, only 0.65 km less than $\sqrt{5}\varphi R/12 = 1023.96$ km. The latter formula has both $\sqrt{5}$ and φ to represent the golden mean, and also contains the important number 12.

I will close this section on Jovis Tholus with 2 examples of triangles involving the eastern centre of the mountain (Fig. 8.9). The first triangle is isosceles in nature with the 2 equal sides emanating from Alba Mons to Jovis Tholus East and to the AscSC1 Crater. These have the sacred distance $\pi R/(3e)$ km which is the same distance from Alba Mons to Jovis Tholus West although in Fig. 8.6 it was put in the form of $2\pi R/(6e)$ to express it in terms of the equatorial circumference of Mars. The equal sides of this isosceles triangle have π/e embedded in their sacred distance formulae whereas the base between Jovis Tholus East and the AscSC1 Crater has e/π in its formula. The distances from Alba Mons and the AscSC1 Crater to Jovis Tholus East are without error since they were used to survey the coordinates of Jovis Tholus East. The distance between Alba Mons and the AscSC1 Crater is 0.54 km greater than its sacred distance formula of $\pi R/(3e)$ km. The term $R/3$ is common to the distance formulae for all 3 sides of the triangle. The second triangle involving the eastern centre of Jovis Tholus is with the Paros Crater and Ascraeus Mons. All 3 sides of this triangle use the northern polar radius (R'; see next chapter) of Mars rather than the equatorial radius (R). The distance from Jovis Tholus East to the Paros Crater is only 0.36 km less than $\varphi R'/5 = 1092.56$ km and to Ascraeus Mons is only 0.74 km greater than $R'/4 = 844.05$ km. The distance between Ascraeus Mons and the Paros Crater is only 0.46 km less than $eR'/(4\pi) = 730.32$ km.

Apollinaris Mons

The final mountain that I examined for this book is Apollinaris Mons. This was by far the hardest mountain to figure out. I examined over 50 craters and other sites to determine if they could provide any clue as to the original centre of this mountain. I came up empty handed. It was exceedingly difficult to find a centre based on supporting evidence because the mountain's perimeter does not offer any good arcs on which to provide an initial estimate of centre. Hence, to be credible, the location

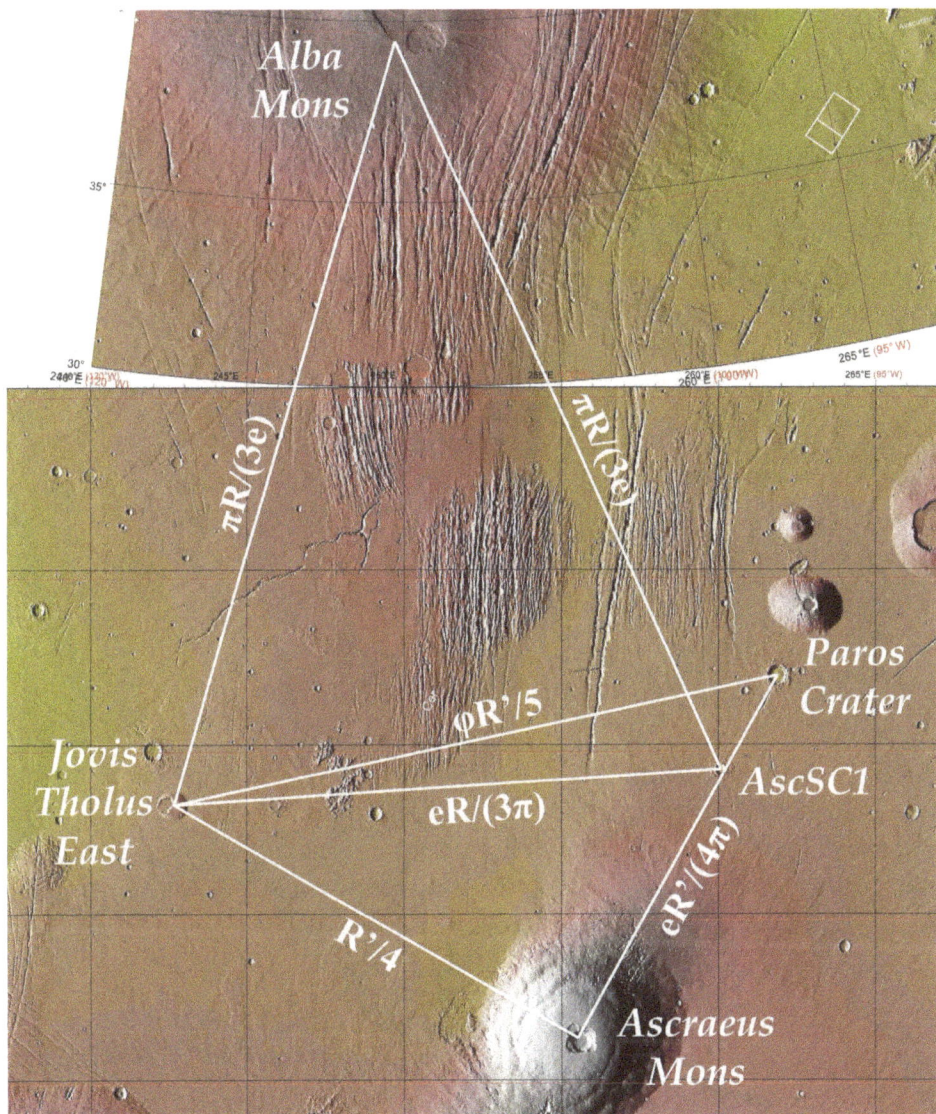

Fig. 8.9: *Triangles involving the eastern centre of Jovis Tholus. The triangle from Alba Mons is isosceles with the ratio π/e in the distance formulae for the 2 equal sides, and the ratio e/π in its base formula. The sides of the triangle to the Paros Crater and Ascraeus Mons all use the polar radius R' in their sacred distance formulae. USGS Astrogeology.*

of the centre had to offer something special (e.g., very low deviations for simple sacred distance formulae to several sites, 3 sites having the same sacred distance to the location, association with a group of sites presenting some kind of unique pattern).

Apollinaris is an extremely old mountain, estimated to be more than 3.5 billion years old[5]. It has long effusive lava flows that extend both to

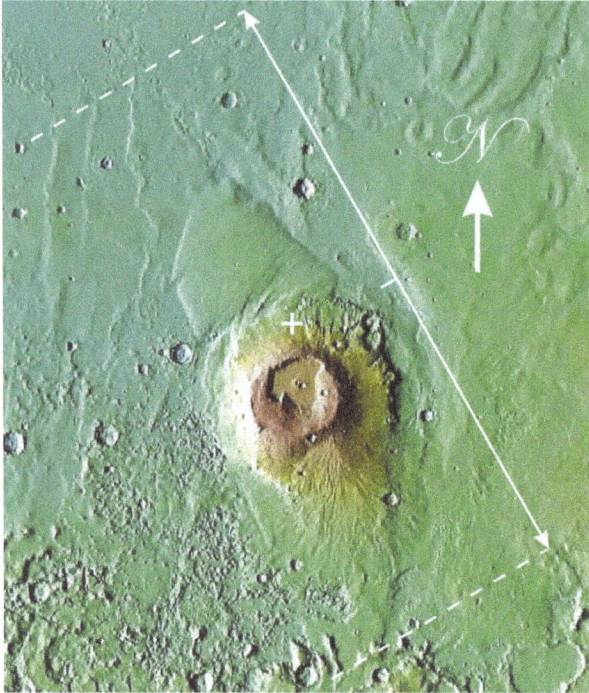

Fig. 8.10: *Survey centre (white cross) of Apollinaris Mons based on being Re/φ km from Issedon Tholus, Uranius Mons and the northern centre of Tharsis Tholus. It lies on the northern slope of the mountain edifice approximately half of the way between the furthest length-wise extent of the lava flow. The halfway point is shown by the short marker line on the long white line drawn between the 2 dashed lines. USGS Astrogeology.*

the northwest and to the southeast. Its main edifice is about 5 km high above the surrounding plains but its elevation is only about 3.16 km, i.e. it is located in an area which is below datum. The main edifice sits mostly south of a point that is equidistant from the farthest reaches of the lava flows (Fig. 8.10). It has basal scarps (cliffs at the edifice perimeter) similar to Olympus Mons but only reaching 1 km high instead of the 6 km observed with the highest mountain on Mars. Using the edges of the lava flows as my guide, I estimated the mountain to be about 200 km wide by about 400 km long.

I was ready to give up but decided to give it one last try. From a mountain centre estimate based on my best effort at fitting a circle to the edifice perimeter, I noticed that 3 important mountains (Issedon Tholus, Uranius Mons and the northern peak of Tharsis Tholus) were at an approximate distance of Re/φ or 5705.57 km. I then set upon determining the unique point that would fit all 3 mountains to this exact distance. I found a location (Fig. 8.10) which was less than 0.5 km in error for each mountain (the distance to Tharsis Tholus North was 0.23 km less, to Uranius Mons, 0.48 km more and to the survey centre of Issedon Tholus, 0.26 km less than the sacred distance of Re/φ km). The coordinates of the location were 174.1293° E 7.3920° S. At first I was quite dubious since the location was about 15 km north of the current caldera's edge. But when I looked at official coordinate determinations such as those of the

European Space Agency (174.6° E 7.2° S) I realized that their estimate was even further north than mine, so my result was not totally lacking in credibility[6].

The likelihood of this estimate for the centre of Apollinaris Mons receives some support from the distance to the southern centre of Tharsis Tholus being 1.45 km less than $\pi\varphi R/3 = 5754.51$ km, to the Ulysses Tholus Caldera, 1.78 km more than $5R/(\varphi e) = 3860.83$ km and to the Nicholson Crater (see below), 0.76 km more than $\sqrt{5}R/(4\sqrt{2}) = 1342.46$ km. I was also encouraged by the finding that the bearings (angles from due north in a clockwise direction) to each of the 3 mountains which were Re/φ km away could be expressed by sacred geometry formulae. Thus the bearing to the northern peak of Tharsis Tholus is only about 1.2 min of a degree more than the angle $\operatorname{atan}(\sqrt{2\pi})$ or 77.315 degrees. The bearing to Uranius Mons is only about 2.8 minutes of a degree more than $e/\sqrt{5}$ radians or 69.65 degrees. Finally, the bearing to Issedon Tholus is only about 2.7 minutes of a degree less than $\operatorname{acos}(\sqrt{2}/\pi)$ or 63.25 degrees.

From Fig. 8.11 it can be seen that the 3 mountains located at Re/φ km away from the survey centre of Apollinaris Mons form sacred triangles with it. Three isosceles triangles are present with 2 being embedded in the third whose base is between Issedon Tholus and the northern peak of Tharsis Tholus. Since all sides from Apollinaris Mons to the 3 distant mountains are equal, the latter are actually on the circumference of a circle centred on Apollinaris Mons with a radius of Re/φ km. These triangles are absolutely enormous! Their heights extend slightly more than one-quarter of the planet's circumference. It appears that the side going from Apollinaris Mons to Issedon Tholus passes through the centre of Olympus Mons, but in actual fact passes about 34 km north of its survey centre. The clockwise bearing angle of the line from Apollinaris Mons to Olympus Mons is about 0.67 degrees greater than to Issedon Tholus.

Nevertheless, the fact that the surveyed centre is so far off the middle portion of the main edifice of Apollinaris Mons, the confidence in the location of its coordinates remains low, and would require further substantiation such as the discovery of a tower or other structure that marks this spot.

The Nicholson and Pettit Craters

The line going from Apollinaris Mons to Uranius Mons in Fig. 8.11 passes about 18 km north of the centre of the large Nicholson Crater (coordinates 195.5360° E 0.2069° N). This was close enough to draw my attention to the crater. I measured the Nicholson Crater to be about 98 kilometers in diameter. The crater has a raised centre which forms a long oval running

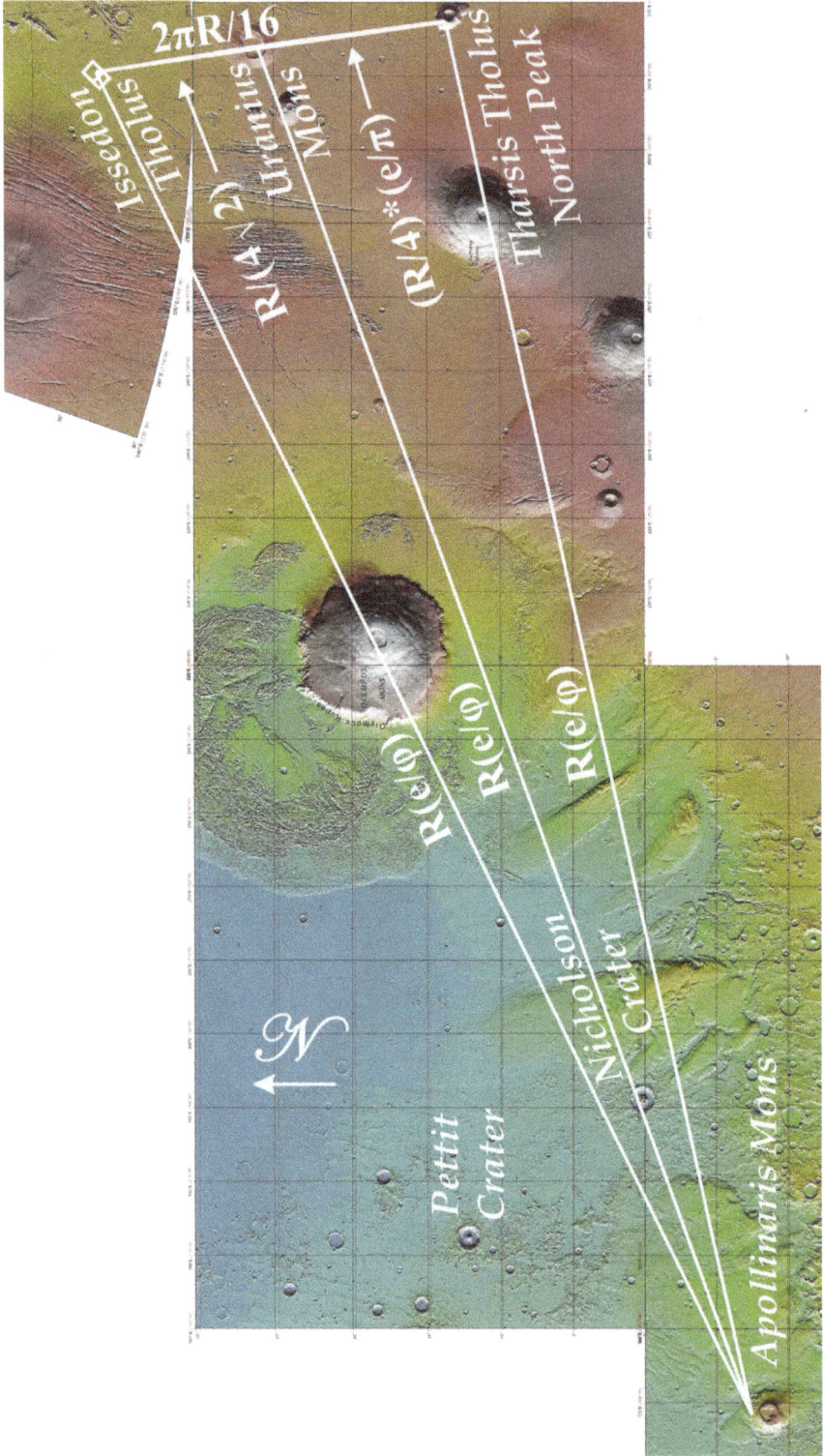

Fig. 8.11: *Three giant sacred isosceles triangles with their apex at the survey centre of Apollinaris Mons. Each equal side is longer than 25% of the planetary circumference at the equator. USGS Astrogeology.*

Table 8.1: *Sacred distances to the Nicholson and Pettit Craters from other sites.*

| Site | Distance to Nicholson/Pettit Crater | | | |
| | Theoretical | | Actual | Differ- |
	Formula	(km)	(km)	ence (km)
Nicholson Crater				
Ayacucho Crater	$3R/\sqrt{5}$	4556.47	4556.36	-0.11
AscSC2 Crater	$(R/2)(\pi/\varphi)$	3297.04	3295.68	-1.36
Olympus Central Caldera	$2R/\pi$	2162.08	2160.48	-1.60
Arsia Mons	$2R/\varphi^2$	2594.46	2595.88	1.42
Biblis Tholus	$7R/\pi^2$	2408.74	2408.83	0.08
Ulysses Tholus	$3R/4$	2547.14	2546.77	-0.37
Jovis Tholus West Centre	$R(e/\pi)$	2938.57	2937.18	-1.39
Apollinaris Caldera	$2R/5$	1358.48	1358.48	0.00
Albor Tholus	$(R/2)(e/\varphi)$	2852.78	2852.66	-0.12
Pettit Crater				
Apollinaris Mons	$\varphi\sqrt{5}R/9$	1365.28	1364.38	-0.90
Olympus Mons	$\varphi\sqrt{3}R/4$	2379.47	2379.28	-0.19
Pavonis Mons	$\sqrt{3}R/\varphi$	3635.51	3635.67	0.16
Ayacucho Crater	$10R/e^2$	4596.24	4596.54	0.30
Olympus Central Caldera	$R/\sqrt{2}$	2401.47	2401.68	0.21
Arsia Mons Caldera	R	3396.19	3395.09	-1.10

exactly north-south. It turns out that the distances of this crater to a number of important mountain and crater sites on the planet can be expressed as simple sacred geometry formulae with very low deviations from theoretical (Table 8.1). The equations in Table 8.1 for the distances to the AscSC2 Crater and Albor Tholus differ only in the use of π/φ or e/φ. The distance equations to Ulysses Tholus and the Apollinaris Caldera use only a ratio of integers multiplied by R. The equation for the distance to the western centre of Jovis Tholus is R multiplied by the familiar ratio of e/π. The distance to the Ayacucho Crater is 3 times the height or base of the bisected isosceles triangle in Chapter 5. The remaining equations are integer multiples of R divided by a single irrational number or by the square of a single irrational number.

Also shown in Fig. 8.11 is the Pettit Crater (186.1673° E 12.2701° N) which is located to the northwest of the Nicholson Crater. Since it seemed

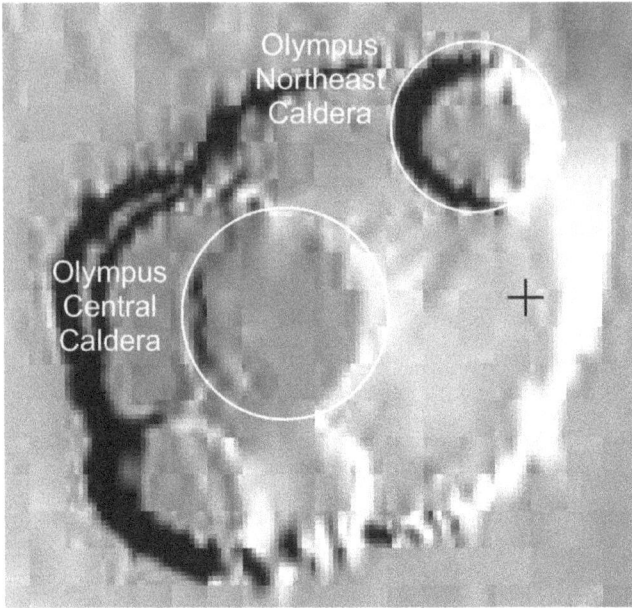

Fig. 8.12: *Caldera complex of Olympus Mons. I named 2 smaller calderas within the complex as the Olympus Central Caldera and the Olympus Northeast Caldera. The survey centre for Olympus Mons is indicated with a black cross. USGS Astrogeology.*

to be about the same size as the Nicholson Crater I decided to examine it as well. The Pettit Crater also has an extensive raised centre and has a diameter of about 90 kilometers. I determined that the location of the crater centre corresponded to a bright spot on the north northeast portion of the raised central structure. This crater was related to Apollinaris Mons by a sacred distance of $\varphi\sqrt{5}R/9$ km (Table 8.1). A sacred distance similar in form is the distance of $\varphi\sqrt{3}R/4$ km to Olympus Mons. The distance to Pavonis Mons is $\sqrt{3}R/\varphi$ km. Note the use of φ and $\sqrt{3}$ in the distances to Olympus Mons and Pavonis Mons. The simplest of all sacred distance formulae is R km to the Arsia Mons Caldera. Also simple is the sacred distance of $R/\sqrt{2}$ km to the Olympus Mons Central Caldera. The distance to the Ayacucho Crater is only 0.30 km more than the sacred distance formula of $10R/e^2$ km.

Distances to both the Nicholson and Pettit Craters are given from the Olympus Central Caldera in Table 8.1. Since the caldera of Olympus Mons is really a large complex of several smaller calderas, I have given names to distinguish 2 of the smaller calderas for which the coordinates of their centres could be easily determined. These are shown in Fig. 8.12 and I named them the Olympus Central Caldera and the Olympus Northeast Caldera.

Conclusion

What started out to be a simple cleanup operation ended up being a long series of remarkable discoveries about how the lesser mountains on Mars relate to each other and to other sites in a sacred geometry layout. While not quite on par with the sacred triangles discovered for Olympus Mons and the Tharsis Montes, and for the Elysium group of mountains, many of the patterns created with these mountains, especially those containing simple distance formulae, strongly suggest that there was an overall plan in their positioning. What is needed now is to sit back and try to gain an overall perspective on all of this information. But before doing this there are still several more amazing discoveries to relate in the next few chapters. Following that, I will provide statistical proof of artificiality and then finally I will try to draw things together.

References

1. *List of mountains on Mars by height. Wikipedia.*
   http://en.wikipedia.org/wiki/List_of_mountains_on_Mars_by_height

2. *Alba Patera, Mars: Topography, structure, and evolution of a unique late Hesperian–early Amazonian shield volcano. Mikhail A. Ivanov and James W. Head. Journal of Geophysical Research, Vol. 111, E09003, doi:10.1029/2005JE002469, 2006.*

3. *Spreading of the Olympus Mons Volcanic Edifice, Mars. P.J. McGovern and J.K. Morgan. Lunar and Planetary Science XXXVI (2005) 2258.pdf. http://www.lpi.usra.edu/meetings/lpsc2005/pdf/2258.pdf*

4. *http://volcano.oregonstate.edu/oldroot/volcanoes/planet_volcano/mars/Cones /jovis.html*

5. *http://www.lpi.usra.edu/publications/slidesets/3dsolarsystem/slide_23.html*

6. *http://www.esa.int/Our_Activities/Space_Science/Mars_Express/Ancient_ caldera_in_Apollinaris_Patera*

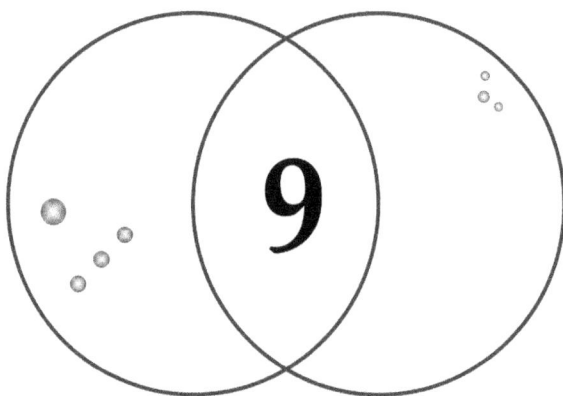

A Pentagram, Poynter and Face

hile searching around for survey crater candidates, I noticed a very peculiar object laying just to the west of Biblis Tholus and to the east of the Gigas Fossae (Fig. 9.1). The object has the definite shape of a 5-pointed star, of which 1 of the points was very clear and the other 4 points only partially visible. Situated to the north of this object are 3 large structures which form a straight line which can be extended to include the survey centre of Pavonis Mons. Several smaller ones are scattered about, some of which are arranged in a second straight line. I discovered that I could overlay the outline of a perfect pentagram (one without the interior pentagon so as to increase visibility of the underlying structure) upon the object provided that I rotated the pentagram counterclockwise by a number of degrees from a configuration that had one of its points aimed due north. However, since the object was incomplete and the resolution was rather low, there was a substantial range of sizes and degrees of rotation that gave credible fits. Other NASA pictures of the region offered little help since they showed a complete absence of a pentagram-shaped structure. This of course gave me cause for great hesitation to treat the object as a pentagram, and I had to wrestle with myself to not simply dismiss it as an aberration of the USGS MOLA maps. Persistence eventually paid off since the relationships of this object to other structures on Mars reinforced its pentagram nature and its extreme importance in the sacred geometry of the Martian mountains.

Fig. 9.1: *A star-shaped object is located just east of the Gigas Fossae (partially shown). The star has 5 points and can be overlaid with a pentagram outline. Only the southwest star point is reasonably complete while the other 4 points are partially present. To the north of the star are 3 large structures arranged in a straight line (bearing angle very close to acos(1/(eπ)) = 83.275°) which can be extended to include the survey centre of Pavonis Mons. Further north of this are several smaller structures also arranged in a straight line with a bearing angle of 60°. The star is rotated counterclockwise from having a point aligned due north. USGS Astrogeology.*

Pentagram Basics

Before I present the study of the star-shaped object, the reader should be familiar with some basic concepts about the geometry and mathematics of pentagrams. This is needed in order to understand how the star-shaped object fits into the sacred geometry of the mountains, and that it does this superbly only if its shape is interpreted to be a perfect pentagram. The pentagram is a 5-pointed star drawn by using only 5 equal straight lines (Fig. 9.2). These lines intersect in such a way that within the interior of the star they form a 5 equal-sided polygon known as a regular pentagon.

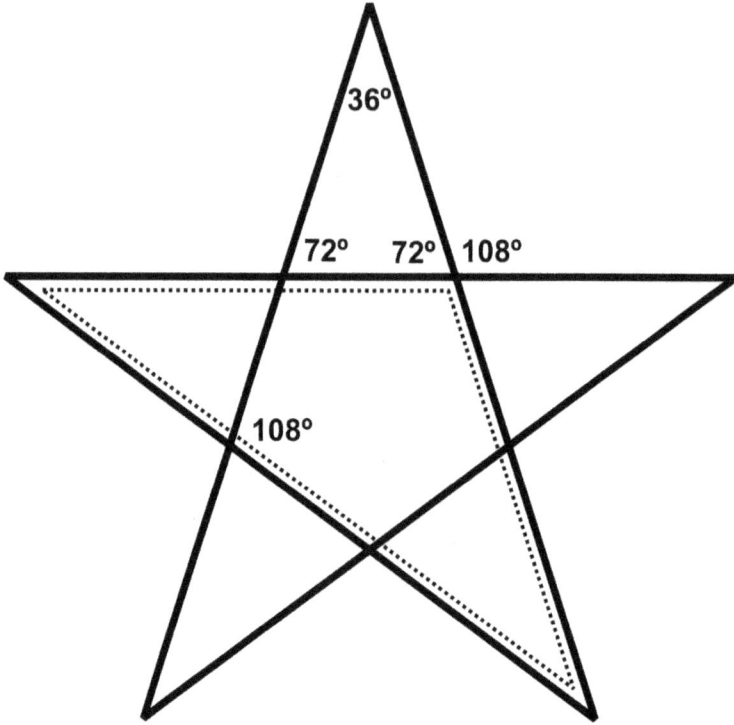

Fig. 9.2: *The pentagram is composed of 5 intersecting straight lines of equal length. Each star point is an isosceles triangle whose equal sides are in golden ratio (1.6180) to its base. The 3 sizes of angle in the pentagram are 36°, 72° and 108°. Dotted lines show an isosceles triangle (golden gnomon) whose base is one of the 5 pentagram lines.*

The pentagram has very interesting mathematical and geometrical properties. It can be considered to contain 10 isosceles triangles (the 5 isosceles triangles of the star points and the 5 isosceles triangles formed by using each of the 5 pentagram lines as a base - see dotted lines in Fig. 9.2). Forming the 5 star points are 10 equal long sides and 5 equal short sides. The ratio of a long side to a short side is exactly equal to φ = 1.6180 thus making each of the star points a special geometric shape known as a golden triangle. A golden triangle is an isosceles triangle in which the smaller side is in golden ratio to either of the larger 2 sides. The ratio of a long side plus a short side to a long side is also equal to φ, as is the ratio of one of the 5 straight lines of the pentagram to the sum of a long side and a short side. Because of the latter ratio, the isosceles triangles formed by using the 5 pentagram straight lines as bases are called golden gnomons which are triangles in which the ratio of each of the shorter equal sides to the long side is equal to φ = 0.6180, the reciprocal of the golden ratio.

Now what about π? Is π represented at all in the pentagram? Actually π is very strongly present in the pentagram since this figure has a circular shape. The tips of the star points are equidistant from the centre of the pentagram and hence are points on an imaginary circle drawn around the periphery of this figure. In fact, this circle is produced in many ancient and modern representations of the pentagram. When the circle is present, the pentagram is sometimes called a pentacle, although very often this term is used synonymously with "pentagram". At any rate, the circumference of this circle divided by the diameter of the pentagram is π.

There are 3 angle values in the pentagram. The angle of each star point is equal to 36 degrees (1/5 of the sum of the 3 angles in a triangle which is 180 degrees) and the angles at the triangle base of a star point are each equal to 72 degrees (2/5 of 180 degrees). The angles of the interior pentagon are all equal to 108 degrees (3/5 of 180 degrees) as are the angles between the points of the star. It would seem that the most important angle here is 36 degrees since all the other angles are integer multiples of it, and it is the angle of the star points. There are also trigonometric equations which are interesting because they show that the 3 angles of the pentagram are intimately related to φ, i.e., the value of φ can be derived from them. These are:

$$\cos(\pi/5) = \cos(36) = \varphi/2$$
$$\cos(2\pi/5) = \cos(72) = 1/(2\varphi)$$
$$\cos(3\pi/5) = \cos(108) = -1/(2\varphi)$$

It should be noted that in these formulae, the symbol π refers to π radians = 180 degrees. By using the radian format for angle units, the 3 major angles of the pentagram can be written as π/5, 2π/5 and 3π/5.

In summary, the following numbers can be used to represent the pentagram or vice versa: φ, 5, 36, 72, and 108. Having seen all of these except 72 in the analysis of Martian mountains and craters so far, it would seem to be prudent to take another close look at the star-shaped object found in the MOLA map of Mars. Also perhaps we should be on the lookout for any number that is a binary factor of the 3 angles of a pentagram. This would include 9, 18, 27, and 54. The value of 18 degrees is obtained by bisecting the star point angle with a line joining a star point to the centre of the pentagram. We have already seen use of the number 9 in Table 8.1, the number 18 in Fig. 8.7 and the number 27 in Fig. 6.3.

Fitting the Pentagram

I made several attempts at fitting this object to a pentagram over the time period that this project took to complete. The fit that I finally settled on draws from several years of experience in analysing the sacred geometry of the topography of the Martian landscape. The most important item that had to be determined was the correct location of its central coordinates since that would result in the greatest likelihood of finding the correct sacred geometry formulae (i.e., the ones that the hypothetical architects intended) for distances to other landforms. The latitude of the centre of the star-shaped object seemed to be slightly less than the value of $\pi°$. Any attempt to bring the overlaid pentagram shape further north to an exact $\pi°$ would position the entire pentagram too far northwards to give a credible fit. I finally hit upon the idea of converting the latitude coordinate from the planetocentric coordinate system to the planetographic coordinate system. If you remember, this latter system takes the ellipsoidal nature of the planet into account instead of assuming it to be a perfect sphere. Suddenly I was very close to a perfect fit! That led me to position the pentagram at exactly $\pi = 3.1416°$ N planetographic coordinates which was determined to be 3.1045° N in planetocentric coordinates.

There is only one other instance of the use of planetographic rather than planetocentric coordinates that I came across on Mars and that is with the placement of the Pavonis Mons survey centre (see Chapter 5). Its latitude was determined to be 1.5993° N planetocentric coordinates. When this latitude is converted to planetographic coordinates it becomes 1.6184° N which is only 1.4 seconds of a degree more than $\varphi = 1.6180°$ N.

Now I had the much more difficult task of determining the most likely longitude coordinate of the star-shaped object. After much experimentation I decided to make the diameter of the pentagram equal to exactly 0.5 latitude degrees (planetocentric) since this size gave an excellent fit to the southwest star point which was the most clearly defined one. Then I had to decide on the amount of rotation the pentagram should be given from having a star point aimed at due north. Although a range of 12 - 18 degrees of rotation in the counterclockwise direction gave credible fits, I finally settled on the value of $\text{atan}(1/(\varphi\sqrt{5}))$ = 15.4504° since it was a sacred geometry formula that gave a strong reference to the golden mean and therefore, the pentagram. This angle of rotation also permitted the northern star point to be perfectly focused on the survey centre of Olympus Mons by adjusting the longitude of the centre of the pentagram outline to 231.6302° E (Fig. 9.3). This is the longitude that I finally settled on for the pentagram structure. With this position, none of the Tharsis Montes line up to any of the star points. It looks like the line to Pavonis Mons from the centre of the pentagram comes close to passing through the junction between the 2 eastern star points, but in reality the bearing angle of the line to Pavonis Mons is about 3° greater than the bearing angle to the junction between the 2 eastern star points.

The fitted pentagram in Fig. 9.3 contains all the bright areas of the star-

Fig. 9.3: *Fitted pentagram contains all the bright areas of the star-shaped object. Its northern star point is aligned to the survey centre of Olympus Mons. The bearing angle of the line to Pavonis Mons is slightly greater than the bearing angle to the junction of the 2 eastern star points. The centre is at π° N latitude (planetographic). USGS Astrogeology.*

shaped object but only the southwest star point is very close to being totally filled up to the edge. The other star points are incompletely filled, especially the northeast star point. The reason for this could be deterioration over a very long period of time or destruction by either a natural disaster or an invading force.

I next looked at a THEMIS image of this region (Fig. 9.4). To my surprise, I found a mound that had planetocentric coordinates that exactly corresponded to the ones I had determined for the centre of the fitted pentagram (231.6302° E 3.1045° N). When I superimposed my model pentagram on the image there was some correspondence of light coloured areas within the pentagram outline but it was far from perfect and was not at all confirmatory of the pentagram shape. There is

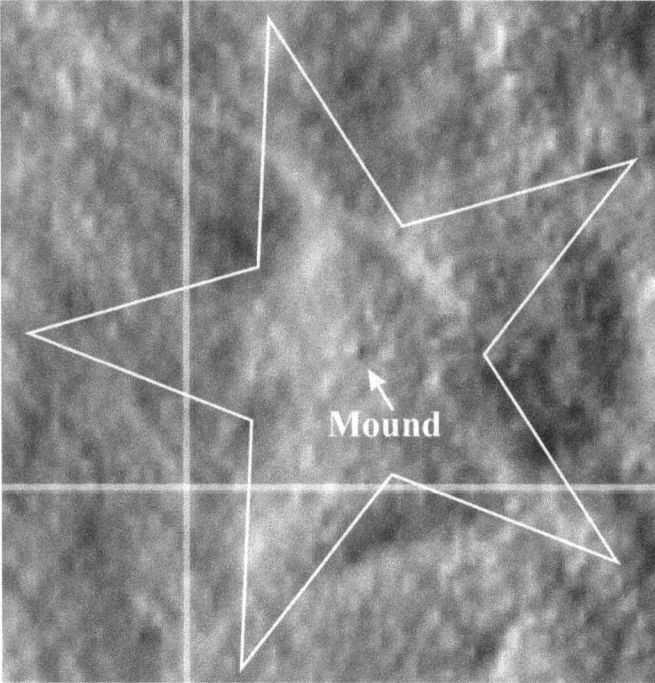

Fig. 9.4: *THEMIS image of the region of the star-shaped object. A mound shares the exact coordinates of the fitted pentagram. Other than this, the image does little to confirm a structure in the shape of a pentagram. USGS Astrogeology.*

therefore a considerable disconnect between the MOLA map image and the THEMIS image of this region.

As a result of the fitting process, we now have a measure of the size of the star-shaped object. The distance from the centre of the fitted pentagram to one of its points is about 14.82 km, which means that a circle drawn around the entire pentagram would have a diameter of about 29.64 km. This is a truly enormous structure, making the great pyramid of Egypt, which has a side length of 0.23 km, look like a child's toy in comparison. An entire city could be placed inside its perimeter. Since the structure seems to have a considerable height, I will henceforth refer to it as the Pentagram Pyramid.

Use of the Polar Radius

Before moving on to measuring the distances between the centre of the star-shaped object and other sites, there is one further concept that I would like to introduce. Except for one of the sides of the lower triangle of Fig. 6.4 and for the 3 sides of the triangle connecting Jovis Tholus East with Ascraeus Mons and the Paros Crater in Fig. 8.9, I have been studiously avoiding using anything but the equatorial radius in sacred distance formulae so as not to overload the reader. But it has become abundantly clear to me that the hypothetical Martian architects most likely used the northern polar radius (which I shall designate as R') in certain sacred distance formulae in addition to the equatorial radius R. Because Mars is an oblate spheroid rather than a perfect sphere, the radius at the poles is slightly shorter than at the equator. It measures out at 3376.20 km at the North Pole rather than 3396.19 km. This doubles the number of sacred distance formulae available for use by the hypothetical architects. There are several interesting examples of the use of the northern polar radius. The distance from the Pettit Crater is $R'e^2/11$ km to the survey centre of Hecates Tholus and $Re^2/11$ km to the centre of the Hecates Tholus Caldera. The distance from the western centre of Jovis Tholus is $\sqrt{3}\varphi R/8$ km to the survey centre of Uranius Tholus and $\sqrt{3}\varphi R'/8$ km to the survey centre of Ceraunius Tholus. The distance from Alba Mons is $R'/3$ km to the survey centre of Ceraunius Tholus and is $R/3$ km to the centre of the Ceraunius Tholus Caldera. The northern survey centre of Tharsis Tholus is $R'/\sqrt{5}$ km from the eastern centre of Jovis Tholus and $R/\sqrt{5}$ km from the western centre of Jovis Tholus. This is the same distance formula for the height and base of the great mountain bisected isosceles triangle (Fig. 5.5) except for the use of the polar radius to Jovis Tholus East. Another simple distance formula from Jovis Tholus East which uses the polar radius is $R'/4$ km to Ascraeus Mons (see Fig. 8.9). All of these distances deviate from theoretical by less than 1 km.

Sacred Distances and Triangles

Returning now to the Pentagram Pyramid, I would like to start by examining its relationships with Biblis Tholus and Ulysses Tholus. Both of these mountains have huge calderas in relation to their size. The caldera for Biblis Tholus actually has 2 centres, a northern one and a southern one, which are about 5 km apart. Ulysses Tholus, in addition to its large caldera, has 2 large craters which I have termed the North (N) Crater and the Southeast (SE) Crater. I had been puzzled by these craters for a long time since they seemed to me to be artificially placed for some purpose. I finally discovered that at least one of their purposes appeared to be to create special distances from the Pentagram Pyramid as you shall see.

Fig. 9.5 shows the distances from Biblis Tholus and its caldera to the Pentagram Pyramid. There are 2 sacred distance formulae deviating less than 1/3 km from the measured value to the survey centre of Biblis Tholus. One of them has the number 18 in the denominator which is half the number of degrees in a pentagram star point. The other formula (uses the polar radius of Mars) has the number 108 in its denominator which is the number of degrees between the star points of a pentagram, or in the angles of the interior pentagon of a pentagram. The distance formula of $eR'/(15\sqrt{5})$

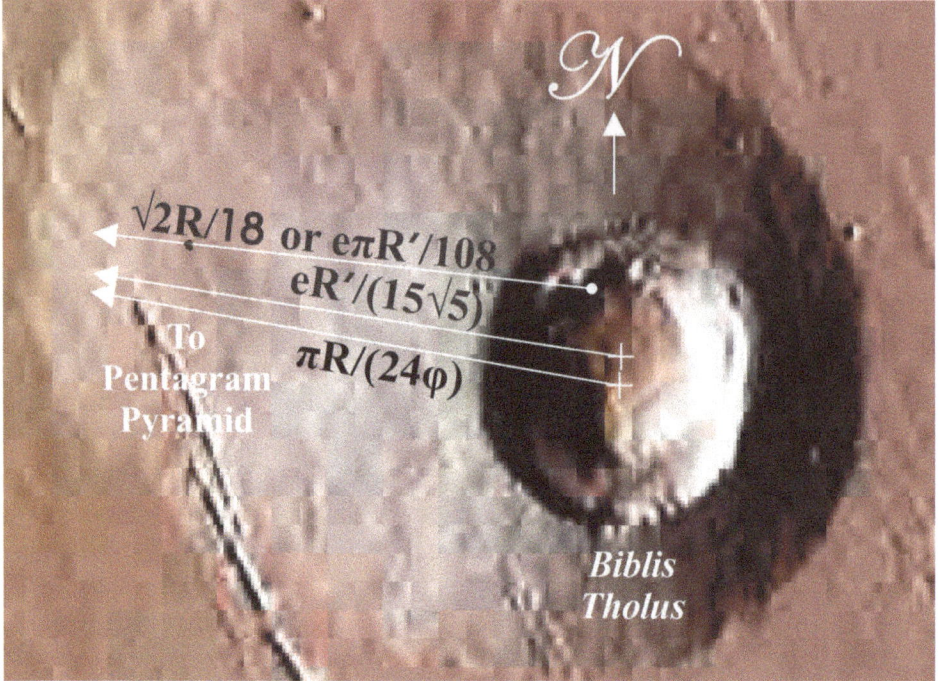

Fig. 9.5: *Distance formulae for Biblis Tholus to the Pentagram Pyramid. Small white dot marks the survey centre for Biblis Tholus. The 2 white crosses mark the north & south caldera centres for which only the closest fitting formulae are given. USGS Astrogeology.*

= 273.62 km fits the north centre of the Biblis Caldera with a deviation of only -0.05 km. Other acceptable formulae with deviations under 1 km are $\varphi R'/20$ km, $\pi R'/(24\varphi)$ km and $\varphi R/(9\sqrt{5})$ km. All 3 formulae contain φ and one contains $\sqrt{5}$, both numbers which relate to the pentagram. The number 9 is 1/4 the angle size of a star point. There are 3 formulae for the distance (274.72 km) from the south centre of the Biblis Caldera to the Pentagram Pyramid which have a deviation less than 0.30 km. They are $\pi R/(24\varphi)$ km, $\varphi R/20$ km, and $e\varphi R'/54$ km. All contain φ which relates to the pentagram. The number 54 is 1/2 the size of the angle between the star points.

The distance formulae from the survey centre, caldera centre and crater centres of Ulysses Tholus to the Pentagram Pyramid are shown in Fig. 9.6. The distance formulae from the 2 crater centres and the caldera centre all have the number 36, the angle of a pentagram star point, in their denominators and all use the polar radius. An alternative formula ($e\sqrt{5}R/50 = 412.86$ km) can be used for the distances from the Ulysses Tholus Caldera and from the north crater of Ulysses Tholus. Both $\sqrt{5}$ and the number 50, which is divisible by 5, could conceivably refer to the pentagram. The distance formula from the survey centre for Ulysses Tholus has the number 72 in its denominator which is the size of the 2

Fig. 9.6: *Distance formulae for Ulysses Tholus to the Pentagram Pyramid. The small white dot marks the survey centre for Ulysses Tholus. The 3 white crosses mark the centres for the north crater, the caldera and the southeast crater of Ulysses Tholus. USGS Astrogeology.*

angles at the base of a star point in a pentagram.

So together, these 2 mountains which lay next to each other refer to all the angles in a pentagram. All of the sacred distance formulae from the Pentagram Pyramid are within 1 km of the measured distances. This offers extremely good evidence for the existence of the Pentagram Pyramid and the validity of my methodology for the determination of the coordinates of its centre. In addition, from Chapter 7 we saw that the distance formula between the survey centres of Biblis Tholus and Ulysses Tholus was $e\varphi R/108$ km. This is analogous to the second distance formula from the Pentagram Pyramid to Biblis Tholus except that the Biblis Tholus to Ulysses Tholus formula uses the equatorial radius instead of the polar radius and the constant φ instead of π.

An examination of the sacred distances to other sites on Mars reveals further connections to the pentagram. In Fig. 9.7 it can be seen that the sacred distance formulae from the Pentagram Pyramid to Olympus Mons, Ascraeus Mons, Arsia Mons and the caldera of Pavonis Mons all have φ in their makeup. You will notice that there are 2 formulae used for the distance to the Pavonis Mons Caldera. The equation $e\varphi R/16$ km matches with $e\varphi R'/10$ km to Ascraeus Mons except for the size of the integer in the

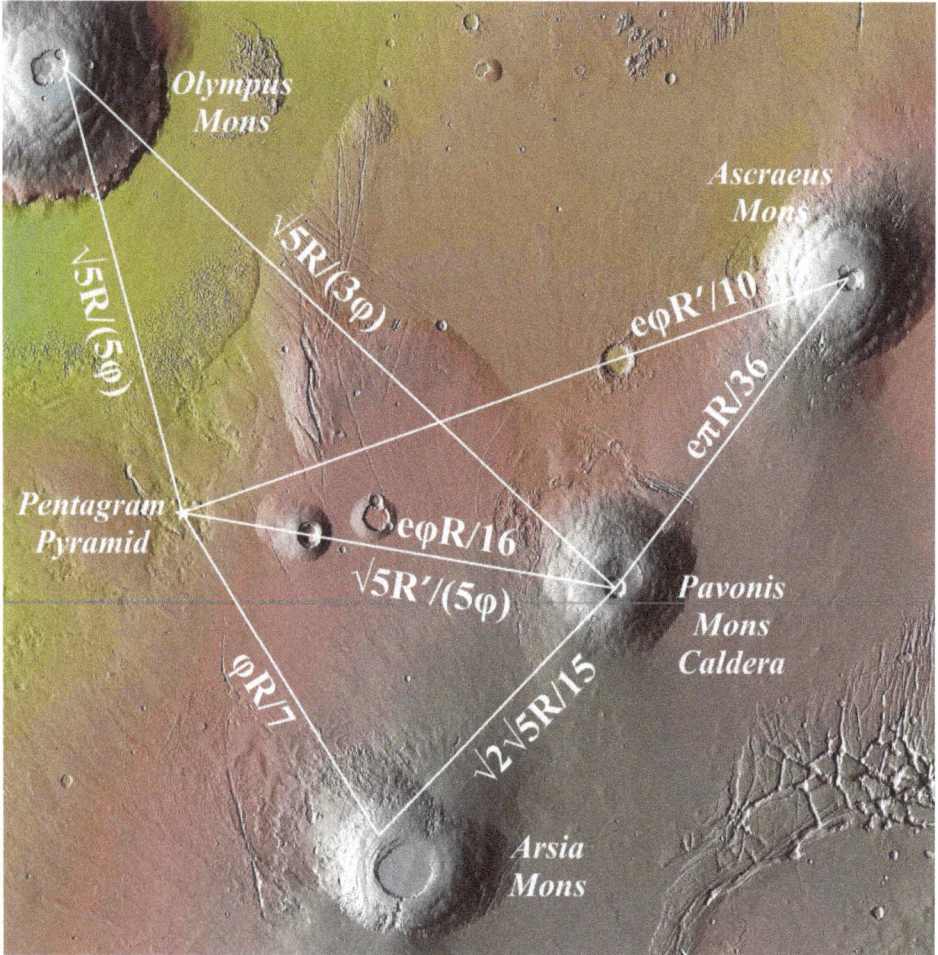

Fig. 9.7: *Sacred triangles emanating from the Pentagram Pyramid. Note how formulae make a matching pair in the distances to Olympus Mons and the Pavonis Mons Caldera, and in the distances to Ascraeus Mons and to the Pavonis Mons Caldera. The 2 formulae to the Pavonis Mons Caldera have almost identical km values. USGS Astrogeology.*

denominator and in the use of R' instead of R. The equation $\sqrt{5}R'/(5\varphi)$ km matches with $\sqrt{5}R/(5\varphi)$ km to Olympus Mons and would result in an isosceles triangle except for the difference between R and R'. Note that the triangle side joining Olympus Mons to the Pavonis Mons Caldera has a distance formula [$\sqrt{5}R/(3\varphi)$ km] which differs from that between the Pentagram Pyramid and Olympus Mons [$\sqrt{5}R/(5\varphi)$ km] only in the size of the integer multiplier in the denominator. The sacred distance from the Pentagram Pyramid to Arsia Mons ($\varphi R/7$ km) has a match in the distance formula to Jovis Tholus West ($\varphi R/5$ km) but I did not show the later in Fig. 9.7 since it would have made the figure too cluttered.

There are so many more sacred distances from the Pentagram Pyramid to other sites that I have elected to show them in Table 9.1 together with those just mentioned rather than try to present them in a narrative or pictorial form. Many of them form groupings of distance formulae which are similar in form, especially eφr/n and eπr/n where r is either the equatorial or polar radius and n is an integer. Hence, I tried to group the entries in Table 9.1 by form of the equation rather than by site name. Some sites have more than 1 formula but I included only some of these to conserve space. If we look at Table 9.1, we see an extraordinary 34 out of 47 sites (sites which I had determined coordinates for so far) with one or more interesting sacred distances to the Pentagram Pyramid that deviated less than 2 km from theoretical distance. All of the giant mountains on the Tharsis Rise plus Olympus Mons are included as well as the 3 mountains from the Elysium group. There are also 8 mountains from the 'lesser' mountains group in the previous 2 chapters, and 8 craters. Now let's look at the sacred distance formulae. The number φ is present in 20 of these formulae, the number √5 in 5 and the number 5 in 3. Incredibly, the sacred distances to Olympus Mons and the Pavonis Mons Caldera contain all 3 of these values. Although it is easy to see how these numbers refer to the pentagram, it is not as readily apparent as to why e would be present in 22 of the formulae and π in 15. The number π is contained in the pentagram circumference so it is not necessarily out of line to find its presence in distance formulae that could represent the pentagram. However, the high frequency of occurrence of the number e in these formulae is mystifying since e does not seem to be measurable in any of the components of the pentagram. I puzzled over this a very long time but never could come up with a solution. It finally occurred to me to look at the intrinsic nature of φ rather than try to find a pentagram component that had the size of e = 2.7183.

The number φ is related to the Fibonacci series of numbers which reflect a constant ratio which nature often uses to expand or grow things. For instance, the number of helices in a pineapple are 8 in one direction and 13 in the other, both of which are Fibonacci numbers. The Fibonacci series consist of a series of integer numbers which are the sum of the previous 2 numbers in the series. Once you are given the starting numbers of 0 and 1, the series can be constructed to infinity. The series begins as follows:

0, 1, 1, 2, 3, 5, 8, 13, 21, 34, 55, 89, 144, 233, 377.....etc.

It is linked to φ because the ratio of any 2 adjacent numbers is an approximation of φ which gets increasingly precise as the series progresses to higher and higher numbers. Already by the time you reach the pair of numbers 233 and 377, you have an approximation of φ accurate to 4 decimal places. In fact, you can get the next number in the

Table 9.1	Distance to Pentagram Pyramid			
		Theoretical	Actual	Difference
Site	Formula	(km)	(km)	(km)
Jovis Tholus West	$\varphi R/5$	1099.03	1097.66	-1.37
Arsia Mons	$\varphi R/7$	785.02	783.94	-1.08
Biblis Caldera S centre	$\varphi R/20$	274.76	274.72	-0.03
Uranius Tholus	$\varphi^2 R/4$	2222.84	2223.26	0.43
Ascraeus Mons	$e\varphi R'/10$	1484.94	1485.39	0.45
Pavonis Mons Caldera	$e\varphi R/16$	933.59	933.44	-0.15
Ulysses Tholus Caldera	$e\varphi R'/36$	412.48	412.98	0.49
Ulysses Tholus N Crater	$e\varphi R'/36$	412.48	412.46	-0.02
Biblis Caldera S centre	$e\varphi R'/54$	274.99	274.72	-0.27
Arsia Mons Caldera	$\sqrt{5}\varphi R'/14$	872.51	872.45	-0.07
Alba Mons	$\sqrt{3}\varphi R'/4$	2365.46	2364.50	-0.97
Jovis Tholus East	$\sqrt{2}\varphi R'/7$	1103.65	1101.84	-1.82
Olympus Mons	$\sqrt{5}R/(5\varphi)$	938.68	937.99	-0.70
Pavonis Mons Caldera	$\sqrt{5}R'/(5\varphi)$	933.16	933.44	0.28
Tharsis Tholus South	$3R'/(e\varphi)$	2302.86	2302.12	-0.74
Hecates Tholus	$4R/(\sqrt{3}\varphi)$	4847.34	4846.87	-0.48
Nicholson Crater	$\pi\varphi R'/8$	2145.24	2145.24	0.00
AscSC1 Crater	$\pi\varphi R'/9$	1906.88	1906.44	-0.44
Ceraunius Tholus	$\pi R'/(3\varphi)$	2185.09	2184.04	-1.05
Biblis Caldera S centre	$\pi R/(24\varphi)$	274.75	274.72	-0.03
Poynting Crater	$eR'/(3\pi)$	973.76	973.82	0.06
Olympus Central Caldera	$\sqrt{3}R'/(2\pi)$	930.70	930.03	-0.67
Elysium Mons	$8R/(\sqrt{3}\pi)$	4993.11	4993.71	0.59
Albor Tholus	$e\pi R'/6$	4805.31	4806.99	1.68
Uranius Tholus Caldera	$e\pi R/13$	2230.97	2229.05	-1.92
Paros Crater	$e\pi R/14$	2071.61	2069.83	-1.78
Olympus NE Caldera	$e\pi R'/30$	961.06	962.50	1.44
Olympus Central Caldera	$e\pi R'/31$	930.06	930.03	-0.03
Arsia Mons	$e\pi R/37$	783.85	783.94	0.08
Ulysses Tholus	$e\pi R/72$	402.81	402.97	0.16
Biblis Tholus	$e\pi R'/108$	266.96	267.15	0.19
Pavonis Mons	$2R/e^2$	919.25	920.72	1.47
Ayacucho Crater	$4R'/(e\sqrt{3})$	2868.36	2867.78	-0.58
Biblis Tholus	$\sqrt{2}R/18$	266.83	267.15	0.32
Issedon Tholus	$\sqrt{5}R/(2\sqrt{2})$	2684.92	2686.52	1.60
Tharsis Tholus Caldera	$e^2 R/11$	2281.33	2282.04	0.71
Ceraunius Tholus Caldera	$eR/(3\sqrt{2})$	2175.96	2176.50	0.55
Fesenkov Crater	$eR'/(2\sqrt{3})$	2649.31	2649.26	-0.05
Biblis Caldera N centre	$eR'/(15\sqrt{5})$	273.62	273.67	0.05
Ulysses Tholus SE Crater	$e\sqrt{3}R'/36$	441.55	440.69	-0.86

series by simply multiplying the current number by φ and rounding the result to an integer. Thus 377φ is 609.999991 which rounds to 610 = 233 + 377. When you get far out in the series, you can take a large Fibonacci number and multiply it by a sequence of powers of φ (where the powers go up by 1) to get successive numbers in the series. To continue with our example, the series proceeds as follows: 377φ, 377φ², 377φ³, 377φ⁴, 377φ⁵.....etc. which round out to 610.00, 987.00, 1597.00, 2584.00, 4181.00 ....etc. and can be replaced with integer values. Thus growth in nature can be considered to occur logarithmically, with the base number equal to φ rather than e or 10. Now all logarithmic systems are interconvertible by simply multiplying the logarithms of one system by a factor to get the logarithms of another system. We are all familiar with multiplying a log_{10} number by the factor 2.303 to convert it to a natural log (abbreviation 'ln') number. To convert a natural log number to a golden ratio log (abbreviation '$log_φ$') number, you just have to multiply the natural log number by $log_φ(e)$ which is equal to 2.0781. To convert a golden ratio log number to a natural log number, simply multiply the golden ratio log number by ln(φ) = 0.4812. We now can create equations to relate e to φ:

$$(1)\ log_φ(x) = 2.0781*ln(x)$$

$$(2)\ ln(x) = 0.4812*log_φ(x)$$

$$(3)\ x = e^{ln(x)} = e^{0.4812*log_φ(x)}$$

Thus any length x of a pentagram component can be expressed as e raised to a power which is formulated in terms of a power of φ (i.e., the $log_φ(x)$ term) multiplied by ln(φ) = 0.4812.

Now we are ready to see how the pentagram represents exponential growth. In Fig. 9.8, the base of a star point is set to 1 arbitrary unit. By doing so, the side of a star point automatically becomes equal to φ. Furthermore, the sum of 1 side and 1 base of a star point becomes equal to φ², and the sum of 2 sides and 1 base of a star point becomes equal to φ³. This forms a neat sequence of powers of φ which can be converted into the series of numbers 1, 2 and 3 by simply taking the $log_φ$ of each of the lengths. With the transformation of these powers into powers of e by multiplying them by ln(φ) = 0.4812 we obtain the series of numbers 0.4812, 0.9624 and 1.4436. Thus when all 5 of the lines of the pentagram are arranged together in pentagram format, each of the 5 original lines is segmented into sections which create a representation of exponential growth which can be expressed either as powers of φ or e. In this way the number e can indeed be used as a strong representation of the pentagram

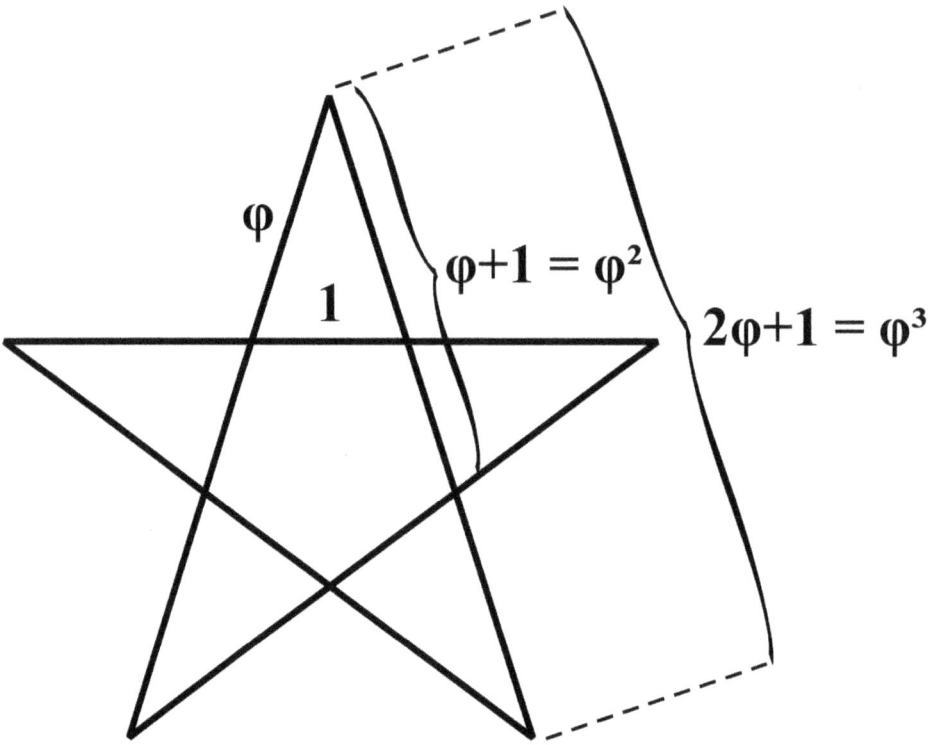

Fig. 9.8: *When the base of a star point is set equal to 1, the other sides of the star point become equal to φ. Then 1 side plus 1 base = φ² and 2 sides plus 1 base = φ³. Thus each of the 5 intersecting lines of the pentagram is symbolic of exponential growth.*

and vice versa.

There is another way in which the pentagram can represent exponential growth. The 5 lines of a pentagram intersect in such a way that in the interior of the star they form a 5 equal-sided polygon known as a regular pentagon (Fig. 9.9a). If you were to connect every second vertex of the interior pentagon with straight lines, another pentagram would be formed within the interior pentagon (Fig. 9.9b). If you were to extend the sides of an exterior pentagon formed by joining the original star points, you would create a larger pentagram whose star points were the intersection of the extended pentagon sides (Fig. 9.9c). Thus the pentagram could be reproduced endlessly in both directions, extending to infinity outwards to ever increasing sizes or shrinking ceaselessly inwards to smaller and smaller proportions. In this sense then, the pentagram can be regarded as a symbol of infinity, and could thus be used to represent the Divine. Interestingly, if you were to divide the 5 original lines by φ^2 you will get the length of the lines for the pentagram which fits inside the interior pentagon. If you multiply the original lines

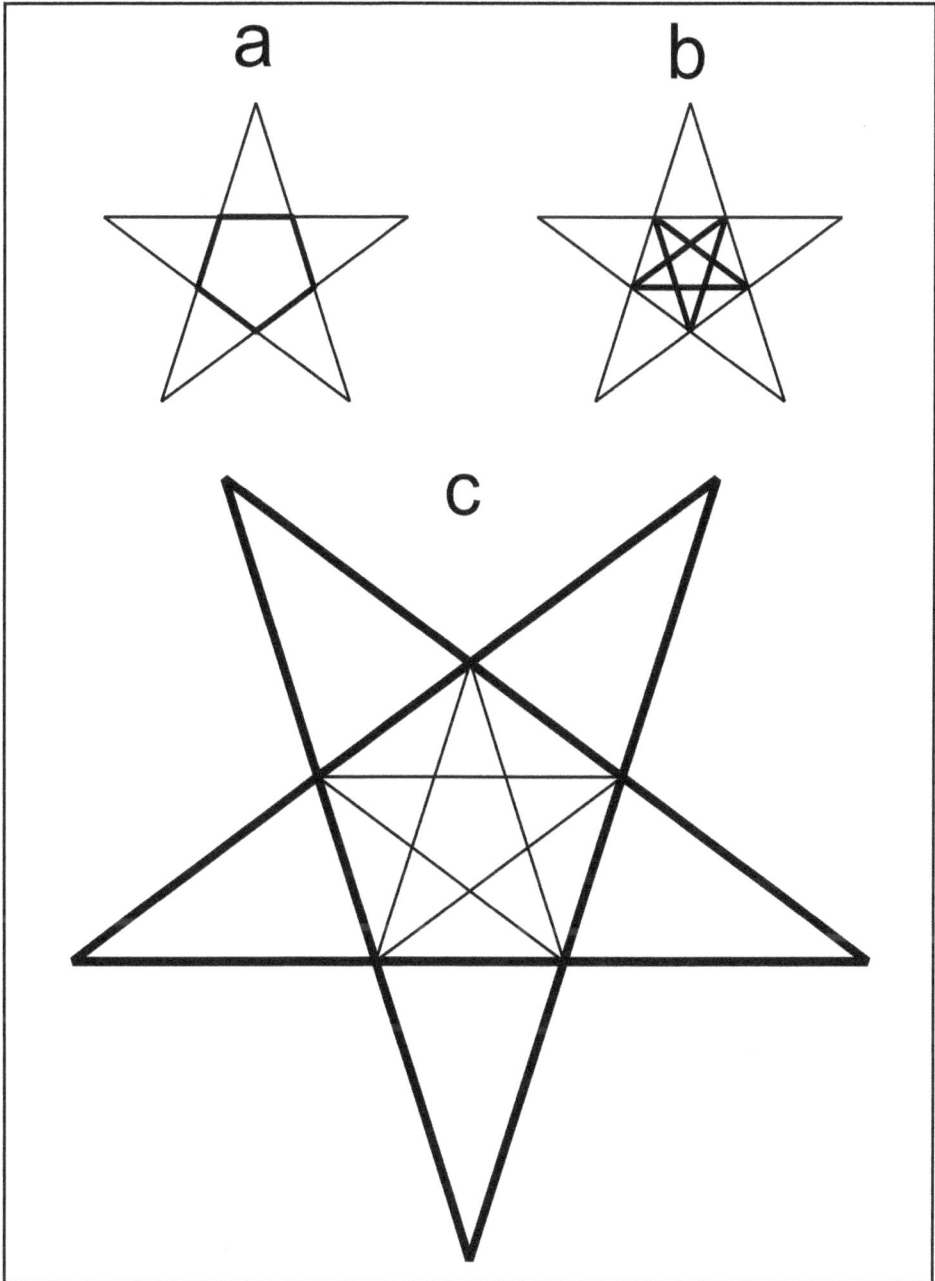

Fig. 9.9: *The pentagram geometrical figure is composed of 5 equal straight lines which create a 5 pointed star with an internal pentagon (a). The pentagram can be reproduced in an infinite series inwards or outwards by dividing or multiplying the lines of the previous figure by φ^2. This is accomplished geometrically by connecting every second vertex of the interior pentagon (b) or by extending lines connecting adjacent points of the original star to where they intersect with each other (c).*

by φ^2 you will get the length of the lines which make up the large pentagram in Fig. 9.9c. By giving the original pentagram (Fig. 9.9a) the same dimensions as the pentagram in Fig. 9.8, the larger pentagram (Fig. 9.9c) will have the following dimensions:

- base of star point = φ^2
- side of star point = φ^3
- 1 side plus 1 base of star point = φ^4
- 2 sides plus 1 base of star point = φ^5

Thus the line segments of Fig. 9.8 which have the values of 1, φ, φ^2 and φ^3 now become φ^2, φ^3, φ^4 and φ^5, i.e., each is multiplied by φ^2. In this way, by reproducing the pentagram endlessly outwards in an expanding series, an infinite series of numbers which are integer powers of φ are created. Since these numbers can also be expressed as powers of e by using equation #3 above, the expanding series of pentagrams can be said to be represented by the symbol e.

We can also construct a formula to calculate the size of the lines of the nth expansion or contraction from the following formula:

$$L' = L\varphi^{2n} = Le^{2n(ln(\varphi))}$$

where n is an integer giving the number of iterations for expansion (positive n) or contraction (negative n), L is the original length of the lines and L' is the new length of the lines. The quantity of 2 * ln(φ) can be considered to be the rate constant of expansion or contraction and it is approximately equal to 0.9624 which is fairly close to 1. Since this formula uses e, the expansion or contraction is in a format which shows very clearly a way in which e can symbolize the pentagram.

There is yet another way in which e can represent the pentagram and that is from the formulae e ~ $42\varphi/25$ = 2.718297 which is accurate to 4 decimal places (e = 2.718282). This approximation of e could be used in situations such as in Table 9.1 where very high accuracy is not required.

The pentagram has now been shown to encode φ, π, and e. It also encodes $\sqrt{5}$. This can be easily shown by taking a pentagram where the long side of a star point is 1 unit long instead of φ units long. The base of a star point then becomes $1/\varphi$ units long and $\sqrt{5}$ is obtained by adding a long side to 2 base sides (i.e., $1 + 1/\varphi + 1/\varphi = \sqrt{5}$). This leaves only $\sqrt{3}$ and $\sqrt{2}$ not encoded by the pentagram and they can be obtained from the equilateral triangle (height = $\sqrt{3}$ for an equilateral triangle of 2 units per side) and the square (diagonal of a unit square = $\sqrt{2}$).

The Poynting Crater

The Poynting Crater sits just north of Pavonis Mons. It is a rather large crater having a diameter of about 67 km. The coordinates of the centre of

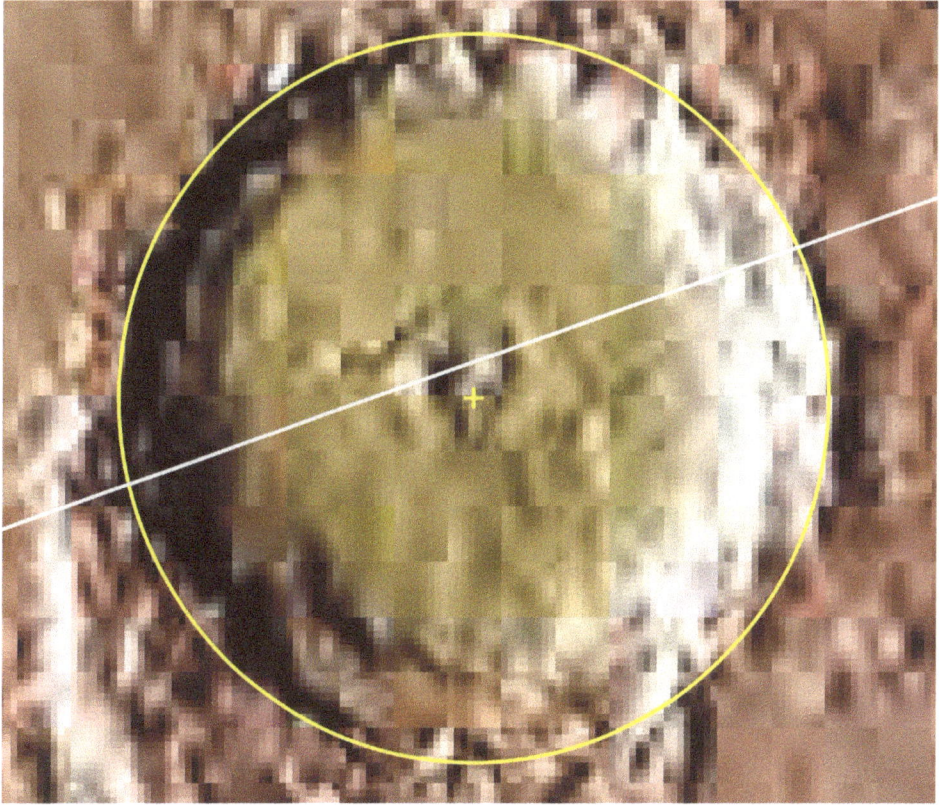

Fig. 9.10: *Poynting Crater with a circle drawn to fit its perimeter. The white line joins the Pentagram Pyramid with the survey centre of Ascraeus Mons. It passes through 2 light coloured regions near the crater centre (yellow cross). USGS Astrogeology.*

a circle which fits its perimeter are 247.2630° E 8.4113° N. These coordinates did not seem to give the Poynting Crater a role as a survey crater since I could not find another crater which would produce a matching sacred distance to the centre of any of the nearby mountains. The crater centre does create some interesting sacred distances, however, such as being only 1.21 km more than the distance of R'/2= 1688.10 km (where R' is the polar radius of Mars) to the Fesenkov Crater and only 0.99 km more than the distance of R'/φ^2 = 1289.59 km to Olympus Mons.

What really drew my attention to the Poynting Crater was the discovery that a line joining the centre of the Pentagram Pyramid to the survey centre for Ascraeus Mons passed over 2 bright regions in the central region of the Poynting Crater but missed the centre of the crater (Fig. 9.10). The western bright region seemed to have the shape of an arch with the line passing through its arms. There seemed to be other bright regions not far from the crater centre so I decided to have a closer look.

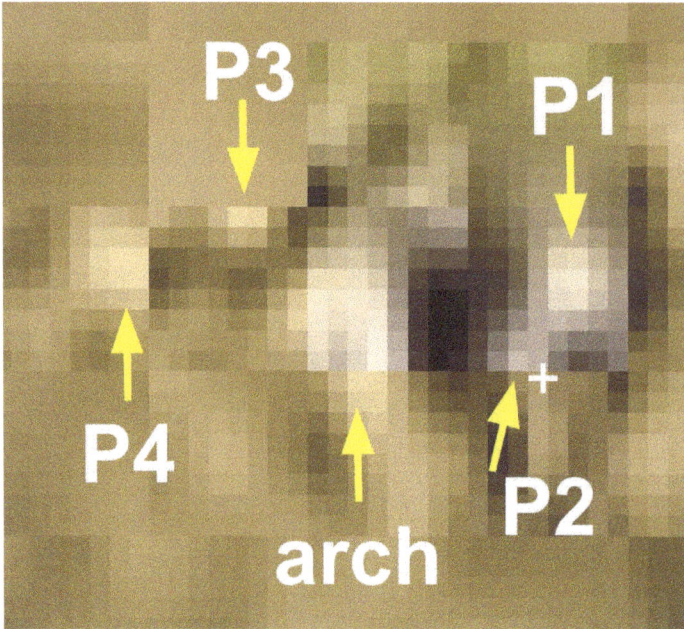

Fig. 9.11: *Central area of the Poynting Crater with 4 bright coloured areas labelled as peaks and 1 larger area which has the shape of an arch and is therefore labelled as such. The centre of the Poynting Crater is indicated by a white cross. USGS Astrogeology.*

With higher magnification I noted that the other regions of brightness were fairly well defined. I labelled the ones in Fig. 9.11 which I found to be of particular interest with the letter P since these regions may be small peaks. I labelled the arch-shaped region as "arch".

What I noticed about these particular bright areas is that they aligned with important sites far beyond the crater perimeter. A line drawn from the centre of the Poynting Crater to the survey centre of Olympus Mons passes through the centres of both the P2 and P3 regions and travels through the northern part of the arch region (Fig. 9.12). A line drawn from the middle of the 3 brightest pixels on the eastern arm of the arch region to the centre of the Ascraeus Mons Caldera passes through the centre of the P1 region whereas it goes slightly north of the P1 centre when drawn to the Ascraeus Mons survey centre. A line drawn from the centre of the P1 region to the survey centre of Arsia Mons passes through the P2 region. Finally, a line drawn from the centre of the P3 region to the centre of the Pentagram Pyramid passes through the middle of the P4 region. Thus the bright regions in the Poynting Crater appear to be aligned to 2 of the giant mountains, to the Ascraeus Mons Caldera and to the Pentagram Pyramid.

However, these are not the only sites that the bright regions are aligned to. If a meridian line is drawn northwards from the survey centre of Pavonis Mons so that it passes through the Poynting Crater, it does so right through the centre of the P3 region (Fig. 9.13). Similarly, if a meridian line is drawn northwards from the centre of the Pavonis Mons Caldera, it runs

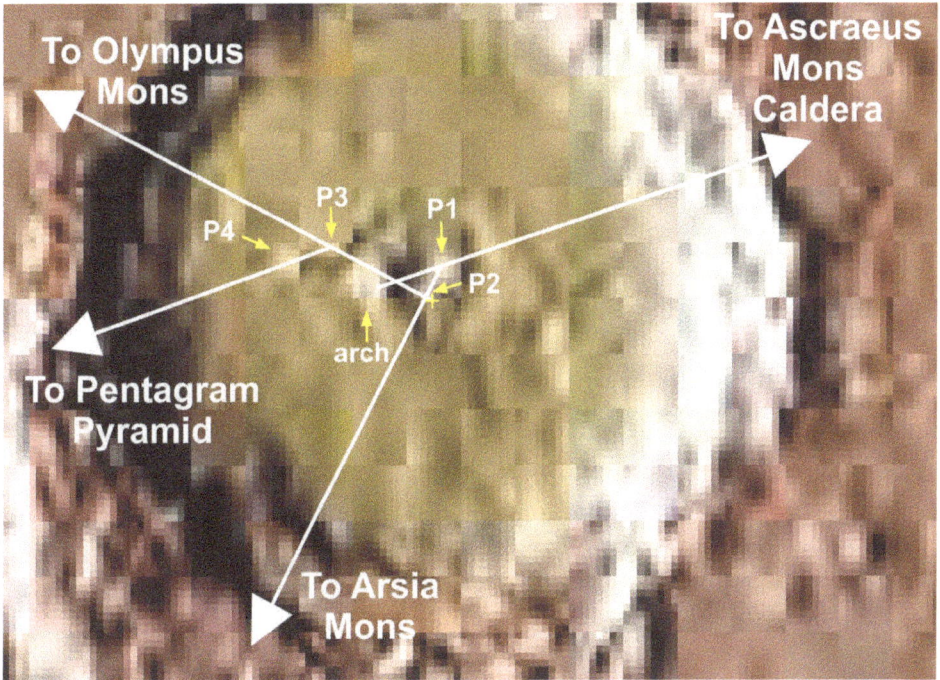

Fig. 9.12: *The bright coloured areas in the Poynting Crater are aligned to important sites far beyond the crater perimeter. P1 and the midpoint of the 3 brightest pixels on the eastern arm of the arch region are aligned to the Ascraeus Mons Caldera. P1 and P2 are aligned to Arsia Mons. The centre of the Poynting Crater, P2 and P3 are aligned to Olympus Mons. P3 and P4 are aligned to the Pentagram Pyramid. USGS Astrogeology.*

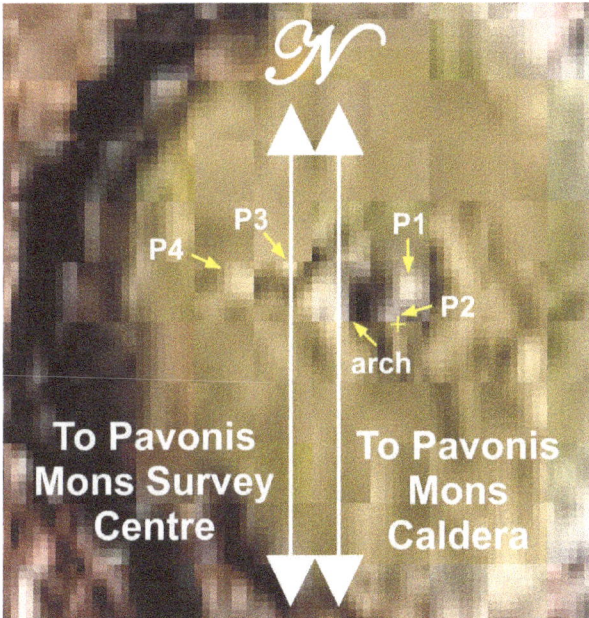

Fig. 9.13: *A meridian line drawn from the survey centre of Pavonis Mons passes through the P3 area of the Poynting Crater. A meridian line drawn from the centre of the Pavonis Mons Caldera runs along the eastern arm of the arch area in the Poynting Crater and is 100.00° E of Elysium Mons. USGS Astrogeology.*

right along the eastern arm of the arch region of the Poynting Crater. What is amazing about this last meridian line is that it is 100.00° E of the survey centre for Elysium Mons! The exact alignment of the current caldera of Pavonis Mons to such a numerically important longitudinal difference from Elysium Mons raises the question of artificiality in the placement of either the survey centre of Elysium Mons or the Pavonis Mons Caldera, or both. It is highly improbable that this could have happened naturally and especially, as we will see in *Intelligent Mars II*, when this type of precise and meaningful longitudinal displacement happens more than once.

What is also remarkable about the 2 meridian lines from Pavonis Mons and its caldera is that they can be used as indicators to due north. Thus the Poynting Crater would make a pointer to the North Pole with reference to either the survey centre for Pavonis Mons or its caldera centre. In addition, with the use of the bright regions, it points to 2 of the giant mountains, the caldera of Ascraeus Mons and to the Pentagram Pyramid. Hmmm.... Poynting.... pointing? Did the person who named the Poynting Crater know something about this? Could it be a deliberate play on words? According to the Wikipedia, the name Poynting was given to the crater in 1988 and it refers to John Henry Poynting who developed the Poynting vector which describes the direction and magnitude of electromagnetic energy flow. Interestingly, he was a professor of physics in the late 19th century at the Mason Science College which is now the University of Birmingham in England. He also has the distinction of having a large crater on the far side of the moon named after him.

There is one other interesting thing that I would like to draw your attention to. The distance between the survey centre of Ascraeus Mons and the middle of the 3 brightest pixels on the eastern arm of the arch region is only 0.03 km greater than the sacred distance of $\sqrt{3}R'/(7\varphi) = 516.30$ km. The same sacred distance is only 0.01 km less than the distance from the survey centre of Ascraeus Mons to the centre of a sizeable crater to the east of the mountain (Fig. 9.14). Hence both the midpoint of the 3 brightest pixels on the eastern arm of the arch in the Poynting Crater and this new crater could be considered to be survey points for Ascraeus Mons in addition to the AscSC1 and AscSC2 survey crater pair. The reason that I missed the new crater on my first attempt at finding survey craters (Chapter 5) was that I had not considered more complex sacred distances or the polar radius yet. I called the new crater the AscSC1a Crater as was discussed in Chapter 8 where it appeared in Fig. 8.8. Although the 2 new survey points appear to lie exactly opposite each other from Ascraeus Mons, their bearing angles to and from the Ascraeus Mons survey centre differ by 0.82 degrees which puts them slightly out of alignment with the survey centre.

There are other notable sacred distances from the midpoint of the 3

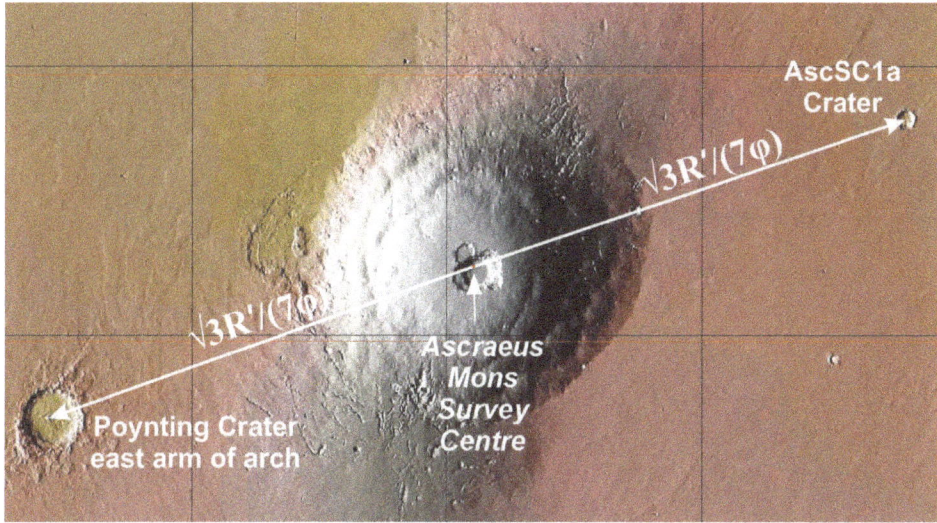

Fig. 9.14: *The midpoint of the 3 brightest pixels on the eastern arm of the arch region in the Poynting Crater acts like a survey point for Ascraeus Mons. It matches the distance of √3R'/(7φ) km from the AscSC1a Crater to Ascraeus Mons. USGS Astrogeology.*

brightest pixels on the eastern arm of the arch. The distance to Uranius Tholus is only 0.44 km more than $2R/5 = 1358.48$ km. The distance to the AscSC1a Crater is 0.32 km more than $3R/\pi^2 = 1032.32$ km and the distance to the Ayacucho Crater is 0.28 km less than $R'/\varphi = 2086.61$ km. Lastly, I will mention that Arsia Mons is 1.91 km more distant than the very significant sacred distance of $\varphi R/5 = 1099.03$ km and that Ulysses Tholus is 0.51 km more distant than $\varphi R/9 = 610.06$ km.

A Huge Face

Next, my attention was drawn to the 3 large objects arranged in linear fashion just to the north of the Pentagram Pyramid. What is particularly interesting is that if you draw a line from the centre of the west-most object to the survey centre for Pavonis Mons, the line passes through the centres of the other 2 structures (turn back to Fig. 9.1). Even more fascinating is that the middle structure has a ghostly image of a face with 2 eyes, a long nose, a mouth, and possibly a beard and headdress (Fig. 9.15, upper picture). The face is perfectly aligned in a north-south direction, and is almost 14 km in length, making it about 5 times the size of the famous face located in the Cydonia region of Mars. The resolution of the face is much too low to see any more detail. The face completely disappears in a THEMIS picture of the same region (Fig. 9.15, lower picture). This is reminiscent of the how the Pentagram Pyramid disappears in the THEMIS photograph except for a small mound to mark its centre.

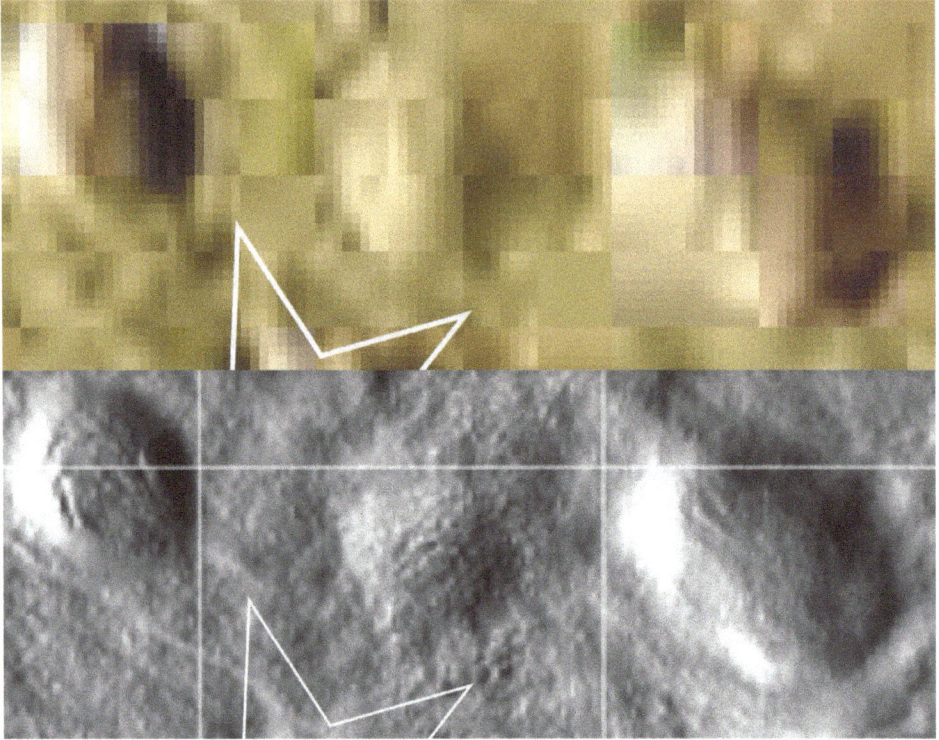

Fig. 9.15: *The 3 structures just north of the Pentagram Pyramid. The middle structure looks like a face in the MOLA map (upper picture) but this disappears in the THEMIS photo of the same region (lower picture). USGS Astrogeology.*

The juxtaposition of a face with the Pentagram Pyramid is very appropriate assuming that the Martian face is proportioned according to the golden mean as has been found for the human face. This face fits nicely inside a golden rectangle (Fig. 9.16) which indicates that the length of the head is close to 1.6180 times its width. It is impossible to determine whether other features of the face are in golden ratio due to the very low resolution of the image. Unbiased high resolution photographs are required to properly establish or dismiss the existence of the face.

Fig. 9.16: *The face fits snuggly inside a golden rectangle suggesting that the length of the face is approximately 1.618 times its width. USGS Astrogeology.*

Summary

This chapter focuses on 3 structures (the Pentagram Pyramid, the Poynting Crater and a Face) which are not mountains themselves but are intimately connected to the major mountains. One of the star points of the Pentagram Pyramid appears to be aimed directly at the survey centre of Olympus Mons. Table 9.1 lists 34 sites having meaningful sacred distances to mountains, mountain calderas and craters from the Pentagram Pyramid. The Poynting Crater has internal structures which align to 2 of the giant mountains and point to the North Pole in combination with Pavonis Mons and its caldera. It also has a site which could be used with the AscSC1a crater to survey the centre of Ascraeus Mons.

Sacred formulae for distances from mountains and calderas to the Pentagram Pyramid seem to emphasize numbers associated with a pentagram, namely φ, π, e and 5. In addition, sacred distances from the Pentagram Pyramid to various features of Ulysses Tholus and Biblis Tholus contain values associated with the size of the 3 angles found in a pentagram, namely 36, 72 and 108. The outline of the face structure has the dimensions of a golden rectangle and is therefore associated with the number φ and the pentagram.

Other pictures of the Gigas Fossae region available over the Internet tend to show a little more detail of the 2 large structures to the east and west of the face. However, both the face and Pentagram Pyramid seem to be more obscured. Is this simply due to differences in shadows and methodology between pictures or could there be something else going on here? Evidence exists that suggests NASA technicians for unknown reasons have deliberately altered certain photos of Mars[1-4]. Whether or not this is the case with the Pentagram Pyramid and face can only be settled by better data.

References

1. *Why did NASA Edit These Images? Ted Twietmeyer.*
   http://www.rense.com/general57/nasa.htm

2. *Update: Did NASA Fake Data To Hide Mars Anomalies? James Burk.*
   MarsNews.com, September 10, 2002.
   http://web.archive.bibalex.org/web/20030625211054/http://www.marsnews
   .com/news/20020910-fakedata1.html

3. *Absolute Proof Of Deception. http://www.checktheevidence.com/articles*
   /Mars-Absolute%20Proof%20Of%20Premeditated%20Lies-adj.htm

4. *Proof That the Face On Mars is Artificial. Tom Van Flandern, Meta*
   Research. [Reprinted from the Meta Research Bulletin 2000/06/15].
   http://metaresearch.org/solar%20system/cydonia/proof_files/proof.asp

10

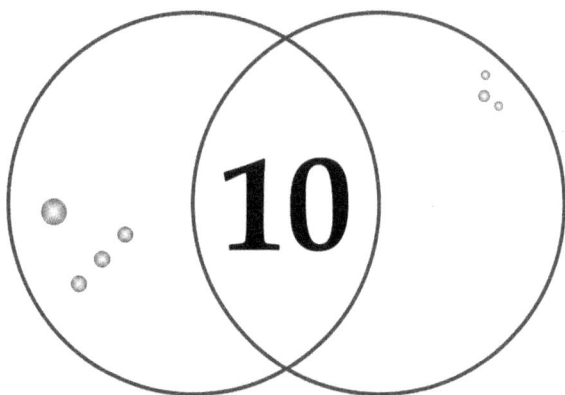

Lessons in Longitudes and Latitudes

I have so far determined the coordinates for 48 major sites including some large craters such as AscSC1, AscSC2, Paros and Fesenkov. The question I would now like to ask is whether or not the longitudes and latitudes of these major sites reveal any interesting relationships to one another or to sacred numbers.

Longitudes and the Prime Meridian

Currently, Martian longitude coordinates are referenced to the Airy-0 Crater which is a 0.5 km diameter crater located within the larger Airy Crater in Sinus Meridiani, a large dark area on Mars situated just south of the equator. Since pixel width is 0.667 km, the Airy-0 Crater is too small to be seen on the MOLA maps. As the Sinus Meridiani location was itself originally designated as the prime meridian in 1830 - 1832 by the German astronomers W. Beer and J.H. Madler, the Airy-0 Crater is hardly likely to have been the prime meridian used by the hypothetical architects of the mountains and craters which have been examined so far. If the sites were located artificially to celebrate principles of sacred geometry, then there is a good chance that they would have been placed at meaningful meridians rather than haphazardly. If I could somehow determine where the original prime meridian was located, I could then examine the longitudes of important sites to see if they revealed special patterns or numbers which conveyed a sacred geometry message. This would also be a test of the artificiality hypothesis since randomly located sites would not be expected to display much sacred geometry in their longitudes.

To carry out my search for a prime meridian, I set the longitudes of all the sites in turn to 0 and examined the longitude values of the other sites when referenced to the new 'prime meridian' candidate. I started with Tharsis Tholus since it is one of the oldest mountains on Mars, with the oldest part being estimated to be 3.82 billion years old. One might expect that a prime meridian would have been selected right from the beginning. When the southern peak of Tharsis Tholus, the most easterly mountain under study, was set to 0° longitude, 4 other sites were found to be an "integer" number of degrees away with an error not exceeding 2 minutes of a degree (± 0.0333°). The northern peak of Tharsis Tholus was 0.97° W, Alba Mons 20.02° W, Arsia Mons 31.02° W and the centre of the Central Caldera of Olympus Mons was 42.97° W. Curiously, there were an additional 7 sites which were very close to being an integer plus 1/2 number of degrees away. These were the Pavonis Mons Caldera at 22.52° W, the Pettit Crater at 83.53° W, the north and south centres of the Biblis Tholus Caldera at 33.49° W, the Olympus Mons Northeast Caldera at 42.49° W, the Tharsis Tholus Caldera at 0.51° W, and Elysium Mons at 122.52° W. The rather extensive use of 1/2 of a degree opens up the intriguing possibility that this civilization may have divided the circle into 720 degrees instead of the 360 degrees that we are familiar with. If you divide 720 by 5, the number of star points in a pentagram, you get 144 which is the square of 12 which we will see in Chapter 12 is a very significant number for Mars. Also 144 is a Fibonacci number and an important number in ancient cultures usually in the form of 144,000, e.g., the number of days in the cycle of time called a Baktun in the Mayan Calendar. The value of 72 degrees or one tenth of 720 is the size of the equal base angles in the 5 isosceles triangles forming the star points of a pentagram.

So right from the start we have a remarkable result: 11 integer or integer + 1/2 values for longitudes in a 360 degree system (I am using the terms "integer" and "integer + 1/2" to describe any longitude within 2 minutes of a degree of an integer or integer + 1/2 value). Could this have happened by chance? Would we get approximately the same result for any site? Apparently not. Setting the northern peak of Tharsis Tholus as the prime meridian yields only 5 integer or integer + 1/2 values in a 360 degree system. Examples of other prime meridian candidates giving low yields were Ulysses Tholus (3 values), Uranius Tholus (2 values), Ceraunius Tholus (3 values), Ceraunius Tholus Caldera (2 values), Poynting Crater (2 values) and the Paros Crater (0 values).

Mathematically speaking, with an error allowance of ± 2 minutes of a degree, there is a probability of 4 min/30 min or 0.133 of being in integer or integer + 1/2 sync with the prime meridian candidate. Since there are 47 test sites in addition to the prime meridian candidate site, then we can expect 0.133 x 47 = 6.27 of the sites on average to be in sync with the prime meridian candidate. When we look at the results from all 48 sites (Table 10.1) we see that the prime meridian candidates have the tendency to be in integer or integer + 1/2 sync with either 5 or with 10 - 11 other sites. Instead of a major peak occurring at 6 synchronized sites, there were only 5 prime meridian candidates which were in sync with 6 other sites. These results are indicative of non-randomness in the longitude coordinates of sites and strongly suggests that the sites have been deliberately positioned by intelligent beings.

A total of 18 prime meridian candidates are in integer or integer+ 1/2 sync with 9 - 12 other sites. These provide the most fertile ground to search for a possible prime meridian. (Of course there is also the possibility that the prime meridian is located elsewhere on the planet and is in integer or integer + 1/2 sync with some 10 - 13 of the sites being examined.) When you look at the sites that are in sync for each of the 18 best candidates, a remarkable discovery emerges. Instead of there being a

Table 10.1. Frequency of Integer or Integer + ½ Values for Prime Meridian Candidates	
No. of Synchronized Sites	**No. of Prime Meridian Candidates**
0	1
1	0
2	3
3	2
4	5
5	9
6	5
7	1
8	4
9	2
10	8
11	7
12	1

single homogeneous group of synchronized sites, there are actually 2 separate groups, suggesting 2 prime meridians instead of one. Each of these groups contain 9 sites that are fully interchangeable with each other, i.e., no matter which one you use as a prime meridian candidate, the other 8 sites will be in sync. The reason that this is less than 10 - 12 synchronized sites is that some sites are near the 2 min error allowance for a given prime meridian candidate and can therefore be outside the limit for another prime meridian candidate.

Assuming that the sites were placed artificially rather than randomly, it is very likely that sites would be located at *meaningful* integer distances from a prime meridian. Hence, one way to spot a prime meridian might be the occurrence of several meaningful integer values in the longitude coordinates of other sites in relation to it. To test this out, I constructed tables for each of the 2 groups of sites in integer or integer + 1/2 sync with each other in their longitude coordinates. Table 10.2 shows the longitude displacements of sites from each of the prime meridian candidates in Group 1. The displacements are in terms of a 360 degree system and are rounded to 1 decimal place. Positive numbers are displacements to the east of the prime meridian candidate and negative numbers are displacements to the west. So far, what we have seen on Mars in terms of meaningful integers are those which are associated with the pentagram. Thus 5, 36, 72 and 108 together with their factors and multiples would be considered meaningful. The only integers in Table 10.2 which might be associated with the pentagram are 24 and 75. The number 24 has 12 as a factor and could therefore be associated with the angles in the pentagram which all have 12 as a factor. The number 75 has 5 as a factor twice, and therefore could be associated with the number of star points in a pentagram. None of these, however, give strong enough

Table 10.2 Longitude displacements (°) from prime meridian candidates for Group 1.

PM Candidate	Apollinaris Mons	Ascraeus Mons Caldera	AscSC2 Crater	Biblis Tholus	Hecates Tholus Caldera	Ulysses Tholus N Crater	Pavonis Mons	Pentagram Pyramid	Ulysses Tholus Caldera
Apollinaris Mons		81.5	75.0	62.0	-24.0	64.5	73.0	57.5	64.5
Ascraeus Mons Caldera	-81.5		-6.5	-19.5	-105.5	-17.0	-8.5	-24.0	-17.0
AscSC2 Crater	-75.0	6.5		-13.0	-99.0	-10.5	-2.0	-17.5	-10.5
Biblis Tholus	-62.0	19.5	13.0		-86.0	2.5	11.0	-4.5	2.5
Hecates Tholus Caldera	24.0	105.5	99.0	86.0		88.5	97.0	81.5	88.5
Ulysses Tholus N Crater	-64.5	17.0	10.5	-2.5	-88.5		8.5	-7.0	0.0
Pavonis Mons	-73.0	8.5	2.0	-11.0	-97.0	-8.5		-15.5	-8.5
Pentagram Pyramid	-57.5	24.0	17.5	4.5	-81.5	7.0	15.5		7.0
Ulysses Tholus Caldera	-64.5	17.0	10.5	-2.5	-88.5	0.0	8.5	-7.0	

evidence to support the candidacy of a prime meridian site in Group 1.

What is very interesting in the second group of sites whose longitudes are in integer or integer + 1/2 sync (Table 10.3) is that of these 9 sites, 4 of them are the mountains positioned at the exact extremes of the Martian mountain ensemble, as if to frame it. Thus Elysium Mons is at the furthest point west, Alba Mons the furthest point north, Arsia Mons the furthest point south, and the southern peak of Tharsis Tholus, the farthest mountain to the east. The integer values in Group 2 are also more interesting than in Group 1. Here we see the integer values of 9, 20, 50, 80, and 100. The number 9 is a factor of the pentagram angles, and all of the other numbers have 5 as a factor. The Olympus Mons NE Caldera produces 3 of these longitude values when used as a prime meridian candidate, and the Biblis Caldera N centre, Elysium Mons and the Pavonis Mons Caldera each produce 2 of these longitude values. From these results, the Olympus Mons NE Caldera emerges as the strongest candidate for the prime meridian. It is also the candidate with the most (12) sites in integer/integer-and-a-half sync.

However, there are other things to consider. For instance, Pavonis Mons has sites in both groups, namely its survey site in Group 1 and its caldera site in Group 2. It is also the central mountain in the Tharsis Montes and its survey site has a latitude of $\varphi°$ N (planetographic coordinates). So these 2 sites of Pavonis Mons offer a central location for the prime meridians of both groups. In the final analysis, however, I decided that the strongest candidate for a prime meridian is the Elysium Mons survey centre. For one thing, it has 2 mountains which are almost exactly $\pi°$ E of it which brings it up to 4 sites with meaningful longitude displacements from it. Now if we add in all the other sites in integer or integer + 1/2 longitude sync, then Elysium Mons would have a grand

Table 10.3   Longitude displacements (°) from prime meridian candidates for Group 2.

PM Candidate	Alba Mons	Arsia Mons	Biblis Caldera N centre	Elysium Mons	Olympus Mons NE Caldera	Pavonis Mons Caldera	Pettit Crater	Tharsis Tholus Caldera	Tharsis Tholus South
Alba Mons		-11.0	-13.5	-102.5	-22.5	-2.5	-63.5	19.5	20.0
Arsia Mons	11.0		-2.5	-91.5	-11.5	8.5	-52.5	30.5	31.0
Biblis Caldera N centre	13.5	2.5		-89.0	-9.0	11.0	-50.0	33.0	33.5
Elysium Mons	102.5	91.5	89.0		80.0	100.0	39.0	122.0	122.5
Olympus Mons NE Caldera	22.5	11.5	9.0	-80.0		20.0	-41.0	42.0	
Pavonis Mons Caldera	2.5	-8.5	-11.0	-100.0	-20.0		-61.0	22.0	22.5
Pettit Crater	63.5	52.5	50.0	-39.0	41.0	61.0		83.0	83.5
Tharsis Tholus Caldera	-19.5	-30.5	-33.0	-122.0	-42.0	-22.0	-83.0		0.5
Tharsis Tholus South	-20.0	-31.0	-33.5		-42.5		-83.5	-0.5	

total of 13 sites with noteworthy longitude displacements from it. The other reason for the suitability of the survey centre of Elysium Mons as a prime meridian is that it is on the far western edge of the region containing all the major mountains of Mars. This allows the longitude coordinates of all the other sites to be expressed in terms of °E. With the other candidates, some of the longitude coordinates would have to be expressed in °E and some in °W to have low valued numbers for longitude values.

Further important discoveries on the prime meridian issue are reported in the second book of this series. Elysium Mons as a prime meridian continued to give strong results for the study material contained in the next 2 books, and is very likely to have been used for this purpose on Mars in the distant past.

Analysis of Latitudes

Having seen that integer and integer + 1/2 values were used to set longitude co-ordinates, or at least, differences in longitude values, it would be reasonable to look for a similar situation with the latitudes of the Martian sites. I was not disappointed, as 9 of the 48 sites showed such a pattern (Table 10.4). Within an error of 2 minutes of a degree, Pavonis Mons Caldera is at 0.5° N, AscSC1a Crater 14° N, AscSC2 Crater 18° N, Paros Crater 22° N, the Ceraunius Tholus Caldera 24° N, Elysium Mons 24.5° N, Uranius Mons 26° N, and the Issedon Tholus survey centre and Caldera both 36° N. All of these sites are at northerly latitudes. Note the series 14, 18, 22, and 26 which progresses by 4 degree jumps and has a difference of 12 degrees between the lowest and highest value. The

Table 10.4: *Sites that have integer or integer + 1/2 latitudes (° N).*

Site	Actual Latitude	Theoretical Integer (+ ½) Value	Difference
Pavonis Mons Caldera	0.5129	0.5000	-0.0129
AscSC1a Crater	14.0093	14.0000	-0.0093
AscSC2 Crater	17.9685	18.0000	0.0315
Paros Crater	21.9949	22.0000	0.0051
Ceraunius Tholus Caldera	23.9674	24.0000	0.0326
Elysium Mons	24.4910	24.5000	0.0090
Uranius Mons	26.0069	26.0000	-0.0069
Issedon Tholus	36.0037	36.0000	-0.0037
Issedon Tholus Caldera	36.0006	36.0000	-0.0006

latitude (36° N) for the Issedon Tholus survey centre and Caldera is equal to the number of degrees in the angle of a star point in a pentagram, and the latitude (18° N) for the AscSC2 Crater is one-half the angle of a pentagram star point.

Upon further investigation, I noticed that a number of latitudes were simple functions of sacred irrational numbers (φ, e, π, $\sqrt{5}$, $\sqrt{3}$, $\sqrt{2}$) which I termed sacred formulae. The simplest sacred formulae (see Table 10.5) were $\varphi°$ N for Pavonis Mons and $\pi°$ N for the Pentagram Pyramid. Both of these latitudes are the only ones measured in planetographic coordinates. The latitudes for the rest of the giant mountains have sacred formulae as well. Thus Arsia Mons is $5\varphi°$ S, Ascraeus Mons is $1/(\pi\varphi)$ radians N, and Olympus Mons is $7(\varphi^2)°$ N. Both Biblis Tholus and the southeast crater on Ulysses Mons have a latitude of $(\varphi^2)°$ N, both the east and west centres of Jovis Tholus have a latitude of $1/\pi$ radians N, and Uranius Tholus together with its caldera and the centre of its round top each have a latitude of $10(\varphi^2)°$ N. The $5\varphi°$ S latitude of Arsia Mons is especially interesting since it honours both the number of star points in a pentagram as well as the role of φ in a pentagram. Other simple latitude formulae are $\varphi/11$ radians N for the Poynting Crater, $4\pi°$ N for Tharsis Tholus South, $7\pi°$ N for the Paros Crater and $9e°$ N for Elysium Mons. Note that the latitude for the Pettit Crater is simply the arcsin of the latitude for the Nicholson Crater. The latitudes of some sites are represented both by an integer or integer + 1/2 number of degrees and by a sacred formula. This happens for the Pavonis Mons Caldera, the Paros Crater and Elysium Mons.

Another very striking feature of the latitudes seen in Table 10.5 is the use of φ values in the triangular region which is bounded by and includes the Pentagram Pyramid and the Tharsis Montes. In this region, the latitude formulae for 9 out of 15 sites contain the value φ. The calderas of Arsia Mons and Ascraeus Mons have $\sqrt{5}$ in their formulae instead of φ. and the Pentagram Pyramid uses π instead of φ. The value of π is contained in the pentagram and the formula $6\varphi^2/5 = 3.1416$ could be used in place of π. And since $\sqrt{5}$ as well as φ is contained in the pentagram, 12 out of the 15 sites could be said to focus on the pentagram in this triangular region. The other sites are the centre of the Ulysses Tholus Caldera, and the north and south centres of the caldera of Biblis Tholus. I could not find any simple sacred latitude formulae for these latter 3 sites.

In all, the latitudes of 37 out of 48, or 77% of the sites are represented by either a sacred formula, an integer or integer + 1/2, or both. This is all the more incredible when you realize that these sites are also positioned in such a way that they create patterns with each other. Hence, more than 1 parameter is being satisfied simultaneously by several site locations.

Table 10.5: *Sites that have sacred latitudes. Latitudes in bold are in planeto-graphic coordinates. Sacred formulae in bold are in radians rather than in degrees.*

Site	Actual Latitude	Formula	Value	Difference
			Sacred Latitude (°)	
Arsia Mons Caldera	-9.3152	-atan(1/(e√5))	-9.3426	-0.0274
Apollinaris Tholus Caldera	-8.5654	-eπ	-8.5397	0.0257
Apollinaris Tholus	-7.3920	-e²	-7.3891	0.0029
Arsia Mons	-8.0996	-5φ	-8.0902	0.0094
Nicholson Crater	0.2069	1/(e√3)	0.2124	0.0055
Pavonis Mons Caldera	0.5129	φ/π	0.5150	0.0021
Pavonis Mons	**1.6184**	φ	1.6180	-0.0004
Biblis Tholus	2.6278	φ²	2.6180	-0.0098
Ulysses Tholus SE Crater	2.6432	φ²	2.6180	-0.0252
Ulysses Tholus	2.9244	**1/(12φ)**	2.9509	0.0265
Pentagram Pyramid	**3.1416**	π	3.1416	0.0000
Ulysses Tholus N Crater	3.5284	**1/(10φ)**	3.5411	0.0127
Poynting Crater	8.4113	**φ/11**	8.4279	0.0166
Ascraeus Mons Caldera	11.1668	5√5	11.1803	0.0135
Ascraeus Mons	11.2791	**1/(πφ)**	11.2715	-0.0076
Pettit Crater	12.2701	asin(1/(e√3))	12.2628	-0.0073
Tharsis Tholus South	12.5955	4π	12.5664	-0.0291
Jovis Tholus East	18.2340	**1/π**	18.2379	0.0039
Jovis Tholus West	18.2598	**1/π**	18.2379	-0.0219
Olympus Mons Central Caldera	18.3211	7φ²	18.3262	0.0052
Olympus Mons	18.3570	7φ²	18.3262	-0.0308
Paros Crater	21.9949	7π	21.9911	-0.0038
Elysium Mons	24.4910	9e	24.4645	-0.0264
Uranius Tholus	26.2088	10φ²	26.1803	-0.0285
Uranius Tholus Caldera	26.1718	10φ²	26.1803	0.0085
Uranius Tholus Round Top	26.1748	10φ²	26.1803	0.0055
Hecates	31.5900	**√3/π**	31.5888	-0.0012
Hecates Tholus Caldera	31.7286	atan(1/φ)	31.7175	-0.0111
Issedon Tholus centre of square	36.1638	10φ√5	36.1803	0.0165
Ayacucho Crater	38.1812	asin(1/φ)	38.1727	-0.0085
Alba Mons	39.4492	4π²	39.4784	0.0292

Arsia Mons as a Prime Latitude

We never use the term prime latitude as it is always assumed that the equator should fulfil this role. However, much to my surprise, I did find a site which marked a latitude which was probably used as an alternative prime latitude to the equator on Mars.

Since Arsia Mons is the most southerly mountain in this group of sites, I decided to calculate the differences in latitudes between this mountain and the other sites (see Table 10.6). First of all, I found that the latitudes of 3 of the other sites were in integer or integer + 1/2 sync with the latitude of Arsia Mons. These were the northern centre of the Biblis Tholus caldera (10.5° north of Arsia Mons), Ulysses Tholus (11°) and the Poynting Crater (16.5°). It was then found that the latitude differences from Arsia Mons of 17 sites could be expressed as sacred formulae. The

Table 10.6: *Integer (or integer + 1/2) and sacred latitude displacements from Arsia Mons. Sacred formulae in bold are in radians rather than in degrees.*

Site	Latitude Difference	Theoretical Integer (+½) Value	Difference
Biblis Tholus Caldera N centre	10.5108	10.5000	-0.0108
Ulysses Tholus	11.0240	11.0000	-0.0240
Poynting Crater	16.5109	16.5000	-0.0109

Site	Latitude Difference	Sacred Latitude (°) from Arsia Mons Formula	Value	Difference
Arsia Mons Caldera	-1.2156	$-e/\sqrt{5}$	-1.2157	-0.0001
Apollinaris Tholus Caldera	-0.4658	$-\sqrt{2}/3$	-0.4714	-0.0056
Apollinaris Tholus	0.7076	$\sqrt{2}/2$	0.7071	-0.0005
Pavonis Mons	9.6989	6φ	9.7082	0.0093
Biblis Tholus Caldera S centre	10.4193	$atan(1/\sqrt{3}\pi))$	10.4134	-0.0059
Biblis Tholus Caldera N centre	10.5108	$\mathbf{1/(\sqrt{3}\pi)}$	10.5296	0.0188
Ulysses Tholus	11.0240	$\mathbf{1/(3\sqrt{3})}$	11.0266	0.0026
Ulysses Tholus Caldera	11.0426	$\mathbf{1/(3\sqrt{3})}$	11.0266	-0.0161
Pentagram Pyramid	11.2041	$5\sqrt{5}$	11.1803	-0.0238
Ulysses Tholus N Crater	11.6280	$\mathbf{\sqrt{5}/11}$	11.6470	0.0190
Poynting Crater	16.5109	$\mathbf{1/(2\sqrt{3})}$	16.5399	0.0289
Olympus Mons NE Caldera	26.8521	$19\sqrt{2}$	26.8701	0.0180
Albor Tholus	26.8933	$19\sqrt{2}$	26.8701	-0.0232
Paros Crater	30.0945	$acos(e/\pi)$	30.0881	-0.0064
Elysium Mons	32.5906	$12e$	32.6194	0.0288
Uranius Mons	34.1065	φ/e	34.1048	-0.0017
Hecates Tholus Caldera	39.8282	$23\sqrt{2}$	39.8372	0.0090

Pentagram Pyramid's latitude is only about 1.4 minutes of a degree farther than a latitude difference of $5\sqrt{5} = 11.18°$ from Arsia Mons. Both 5 and $\sqrt{5}$ are of course highly representative of the geometry of the pentagram. Pavonis Mons was only 0.6 minute of a degree less than $6\varphi°$ and Ascraeus Mons 2.26 minutes (slightly outside my acceptable error of 2 minutes of a degree so I did not include it in Table 10.6) less than $12\varphi°$ north of Arsia Mons. Albor Tholus and the NE Caldera of Olympus Mons were $19\sqrt{2}°$ and Elysium Mons $12e°$ from Arsia Mons. The value for Elysium Mons was especially interesting since it is listed in Table 10.5 to be $9e°$ N from the equator. This means that Arsia Mons must be $3e°$ south of the equator. It is indeed, but with an error of 3.31 minutes of a degree. The use of 12 in both latitude displacements from Arsia Mons for Elysium Mons ($12e°$) and for Ascraeus Mons ($12\varphi°$) is once again a reminder of the importance of this number on Mars which we will see in Chapter 12. Both Ulysses Tholus and its caldera were found to be $1/(3\sqrt{3})$ radians from Arsia Mons. This is a remarkable number since if you can recall, Ulysses Tholus is $R/(3\sqrt{3})$ km from Arsia Mons - an amazing parallel! But it doesn't stop there. Both the north and south centres of the Biblis Tholus Caldera have the latitude difference value from Arsia Mons of $1/(\sqrt{3}\pi)$ in their sacred formulae. However, with the south centre, it is $\text{atan}(1/(\sqrt{3}\pi))$ degrees and with the north centre it is $1/(\sqrt{3}\pi)$ expressed in radians rather than degrees. So once again this mountain, like Ulysses Tholus, is associated with $\sqrt{3}$ and once again, we see the work of genius.

Note that all 3 of the sites which have integer or integer + 1/2 latitude displacements from Arsia Mons also have sacred formulae displacements. Also very notable is the pattern of $0\varphi°$, $6\varphi°$ and $12\varphi°$ for the latitude displacements of the Tharsis Montes (Arsia Mons, Pavonis Mons and Ascraeus Mons) which is an interesting way to link them to the pentagram, very much like the triangular region formed by the Pentagram Pyramid and the 3 Tharsis Montes being populated with many sites having latitude formulae containing φ or $\sqrt{5}$ mentioned above.

Although 17 sites showing notable latitude displacements from Arsia Mons is not as extensive as the 37 sites with notable latitude displacements from the equator, there is sufficient evidence to consider Arsia Mons as a strong candidate for a prime latitude. I will henceforth refer to it as the Arsia Mons Prime Latitude (AMPL). Altogether, with latitudes measured from both the equator and the AMPL, there are 42 out of 48 sites (87.5%) showing notable latitudes, which is truly remarkable. The remaining 6 sites have notable longitude or latitude coordinates in degree systems which are beyond the scope of this book. It is difficult to believe that such sophistication could be simply due to chance.

Summary

It has been discovered that there is a remarkable positioning of a large number of the various major sites at meaningful coordinates. When Elysium Mons (or one of several other sites) is set as the prime meridian, many site longitudes are found to be integer values in what seems to be a 720 degree measuring system. In contrast, latitudes are set to reflect mostly simple functions of sacred irrational numbers in units of degrees or radians, although some are also set to integer values in a 720 degree system. Unexpectedly, Arsia Mons appears to have been used as a prime latitude in addition to the more obvious equator. When you consider that all of these sites are also involved in the creation of sophisticated patterns with each other, the obvious conclusion you arrive at is that there must have been a race of beings with a very high level of intelligence who orchestrated everything in this way. Chance occurrence does not seem to be a credible option. But can this be proven objectively? Such a question will be examined in detail in Chapter 13.

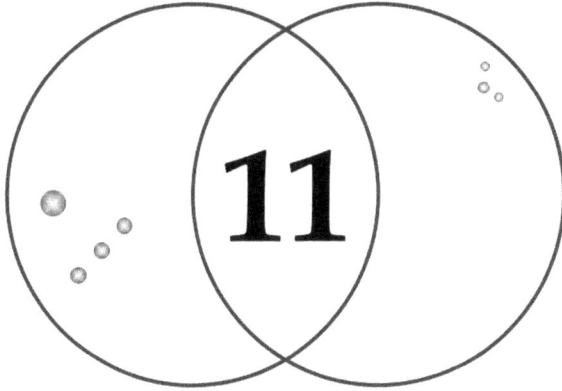

Mountains Bearing Angles

In the first chapter, I mentioned the phenomenon of "inattentional blindness". I must confess that I was a victim of this psychological process during a period of approximately 3 years while studying the mountains of Mars. I was so focused on finding the location of their original centres that I neglected to pay any attention to details about the shape and structural features of these gigantic objects. Yes, I noted such things as the huge multi-kilometer high escarpments of Olympus Mons, the difference between shield volcanoes and tholi, and the vast expanse of Alba Mons, but I did not bother to examine other important aspects of the mountains.

Finally, I happened to glance at Olympus Mons at high magnification one day. I was astonished to see several straight line segments at the outer edge of the mountain. I quickly measured the bearing angles of these lines and found that many of them proved to be angles commonly used in construction, such as 45 and 60 degrees. Wow! Here we have the highest mountain in the solar system, not only placed at a specific location, but also appearing to have been artificially shaped with great precision over enormous distances and heights. What awesome technology could have produced such a structure? And what possible purpose could these architectural features have? Note that, in this chapter, I will use the term 'bearing angle' to refer to angles from due north in either the clockwise (negative numbers) or counterclockwise (positive numbers) direction.

Olympus Mons

I set about measuring the bearing angles of all the straight line segments that I could find on Olympus Mons, either at the outer edge or on the upper surface of the mountain. A sample of these is shown in Figs. 11.1 and 11.2 for the northwest region of the mountain. In Fig. 11.1, the straight lines, one of which extends for about 110 km, mark the outside edges of the mountain top whereas in Fig. 11.2, they mark structures on the cliffs and at the bottom edge of a cliff forming the boundary of the mountain. All of the bearing angles of these lines are multiples of 3, and

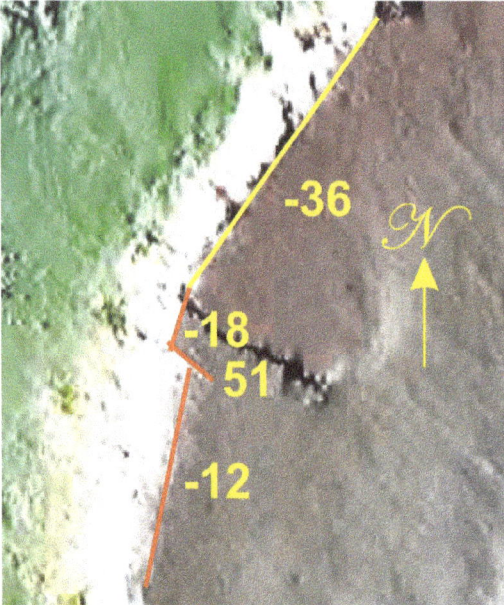

Fig. 11.1: *Northwest side of Olympus Mons showing several straight line edges defining the boundaries of the mountain top. In this and subsequent figures, numbers beside each line are the values for the bearing angles in degrees either in the counterclockwise direction (positive numbers) or clockwise direction (negative numbers). The colour red was simply used to contrast the background. The length of the line with a bearing angle of -36° is 110 km. USGS Astrogeology.*

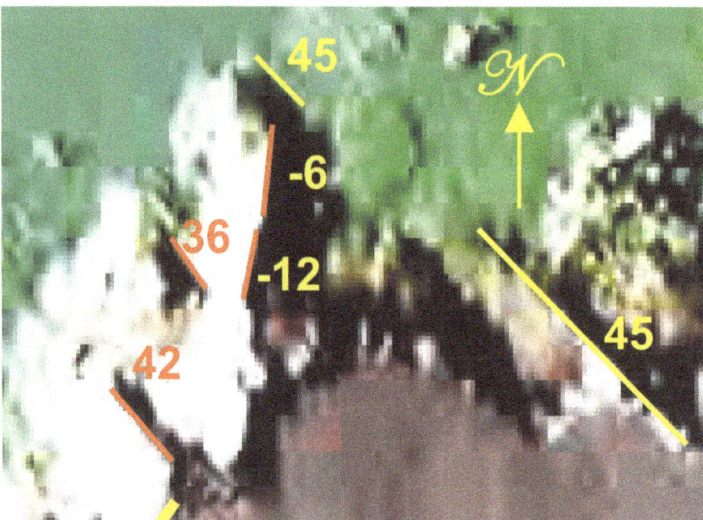

Fig. 11.2: *West part of the Olympus Mons north side showing several straight line edges in the sides of the cliffs forming the mountain border. The short line with a bearing angle of 45° marks the bottom edge of a cliff. The long line with a bearing angle of 45° is 29 km long. USGS Astrogeology.*

some are multiples of 6.

As we work our way around the mountain in a clockwise direction we can see many more examples of straight lines in the mountain structure. Fig. 11.3 shows the east part of the most northerly portion of the mountain. Here, there are more examples of straight lines appearing at the base of cliffs and within cliffs. There is also a straight line bordering the edge of a fracture in the mountain top at its northern edge. It has a bearing angle of 24° in the clockwise direction. In Fig. 11.4 the mountain border starts to turn southward, and a portion of the top edge of the cliff is found to run in a straight line for about 70 km with a bearing angle of 60° in the

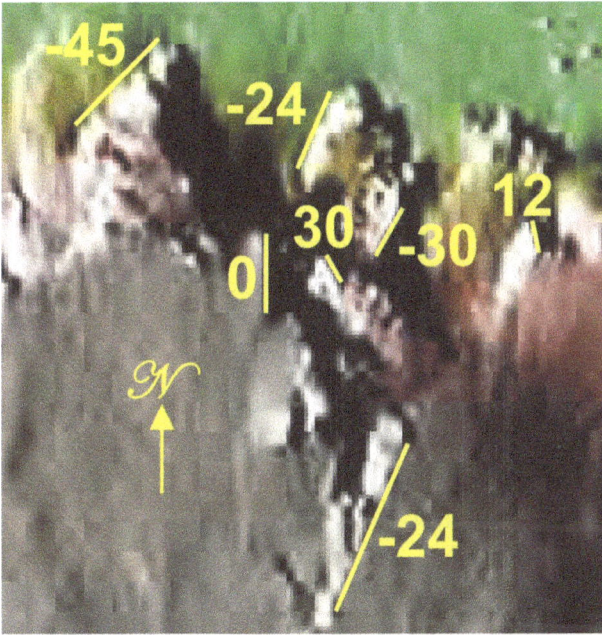

Fig. 11.3: *East part of the Olympus Mons north side showing several straight line edges in the bottom edges and sides of the cliffs forming the mountain border. The edge of a large crevice on the mountain top has a bearing angle of 24° in the clockwise direction. USGS Astrogeology.*

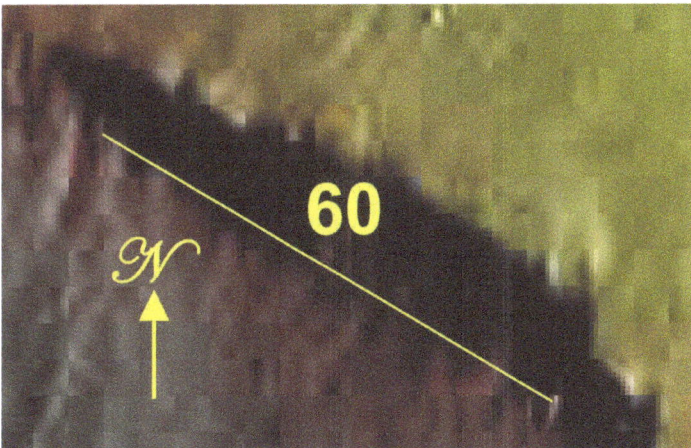

Fig. 11.4: *The edge of the top surface of Olympus Mons in the northeast corner has a bearing angle of 60°. This straight line section stretches for about 70 km. USGS Astrogeology.*

counterclockwise direction. The east side of the mountain (Fig. 11.5) has straight lines mostly at the bottom edges of cliffs, but there is a linear section of the top edge of the cliff as well with a bearing angle of 0°. The southeast corner of the mountain (Fig. 11.6) also has long and short straight line segments at the bottom edges of cliffs. Some of these lines have bearing angles which are the same size as angles found in the star point of a pentagram (-36°, -72°). The bearing angle of 54° is equal to one-half the angle between 2 adjacent pentagram star points. One of the straight lines

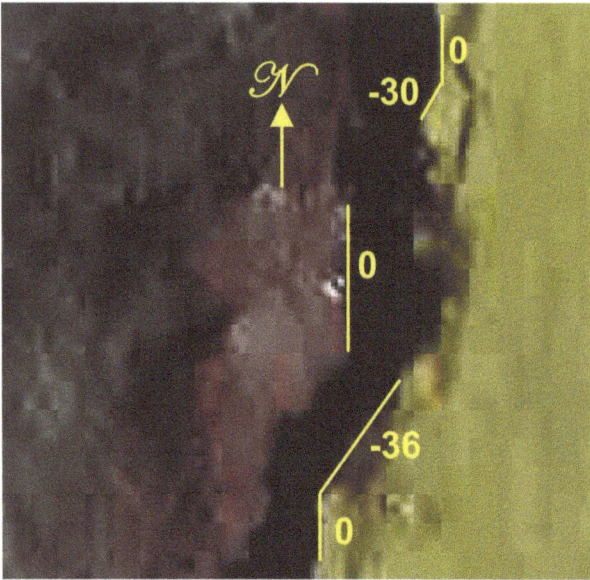

Fig. 11.5: *The east side of Olympus Mons has a linear segment of the top edge with a bearing angle of 0°. Other straight line segments occur at the bottom edges of the cliffs and have bearing angles of 0, -30 and -36 degrees. USGS Astrogeology.*

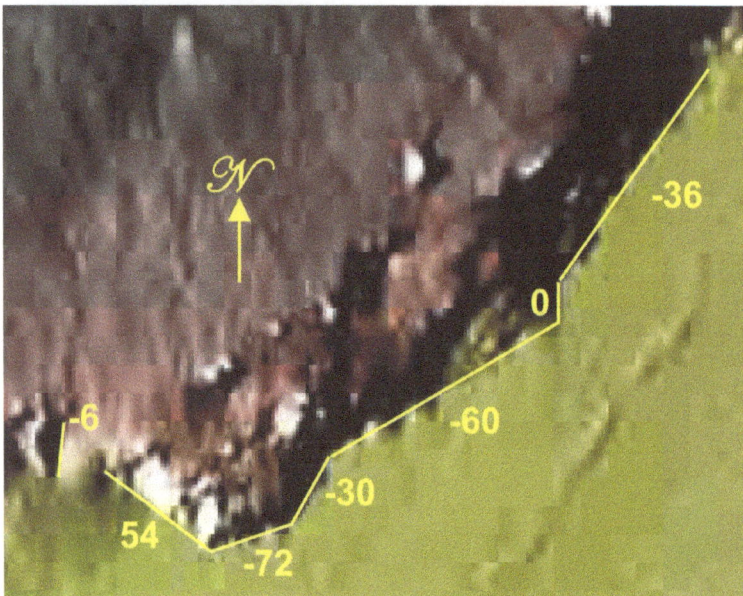

Fig. 11.6: *Straight line segments on the southeast side of Olympus Mons occur at the bottom edges of cliffs. There is also a straight line segment with a bearing angle of -6° which goes up the side of a cliff (bottom left). USGS Astrogeology.*

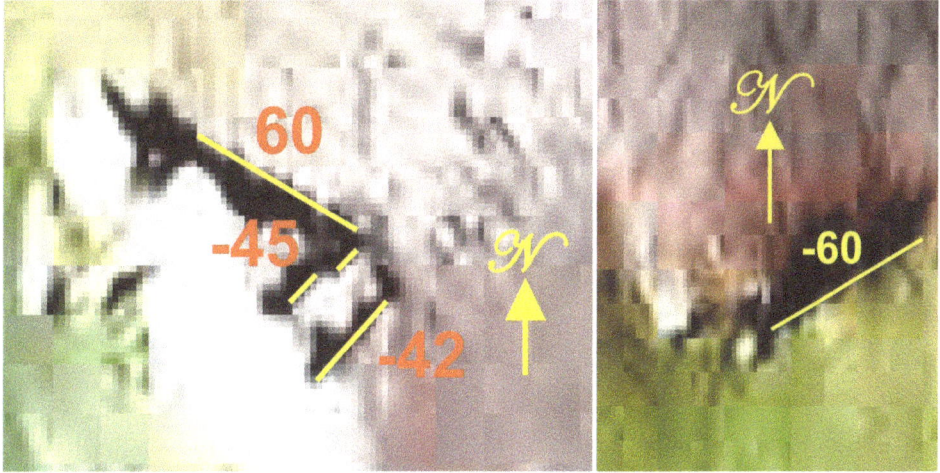

Fig. 11.7: *The west part (left figure) of the south side of Olympus Mons has a deep crevice whose northern edge has a bearing angle of 60°. Other deep crevices in this region have straight line segments with bearing angles of 42° and 45° in the clockwise direction. The middle of the south side of Olympus Mons (right figure) has a straight line segment at the bottom edge of the cliff with a bearing angle of -60°. USGS Astrogeology.*

goes up the side of cliff at a bearing angle of 6° in the clockwise direction.

In the middle and western regions of the southern side of Olympus Mons (Fig. 11.7) there are matching bearing angles of 60° where the one in the middle region runs clockwise at the bottom edge of the cliff and the one in the west region runs counterclockwise along the northern edge of a deep crevice in the side of a cliff. Other edges of this crevice were found to have bearing angles of 42° and 45° in the clockwise direction.

That covers the perimeter of the mountain. Now I would like to turn your attention to a curious triangular region which is present on the top edge of the mountain near the eastern cliffs (Fig. 11.8). Although the

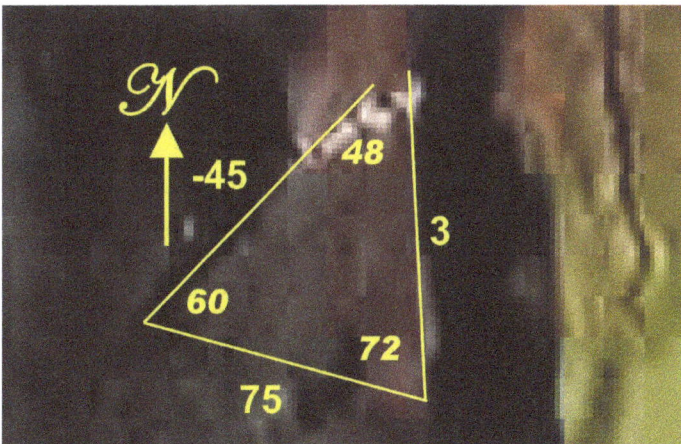

Fig. 11.8: *Triangular region near the edge of the eastern top of Olympus Mons. The numbers in italics are the internal angles of the triangle, not bearing angles. USGS Astrogeology.*

edges are not complete, I was able to use the available information to construct the triangle and measure the bearing angles of its sides (3°, 75° and -45°) and its internal angles (60°, 72° and 48°). Interestingly, all are divisible by 3.

When we look at the caldera region of Olympus Mons, we see that there are several overlapping calderas (Fig. 11.9). I found many of them to have straight line segments in their perimeters which is not what you would expect in structures which are supposed to be round. Many of the shorter straight lines have a bearing angle of 0°. All of the non-zero bearing angles are divisible by 6 except for the straight line segment on the western perimeter which has a bearing angle of about 7.5°.

In summary, the non-zero bearing angles of straight line segments found on Olympus Mons are all divisible by 3 with one exception - the western edge of the caldera complex has a bearing angle of 7.5°. This latter value will show up again with the groove lines discussed in *Intelligent Mars III*. The most frequently occurring bearing angles are 0° (12 values), 45° (6 values), 60° (4 values) and 30° (4 values). The value of 0° creates a pointer to the planetary poles. The values of 60° and 30° might refer to the equilateral triangle since its internal angles are 60°, and 30° is one-half of 60°. The value of 45° could refer to the square or

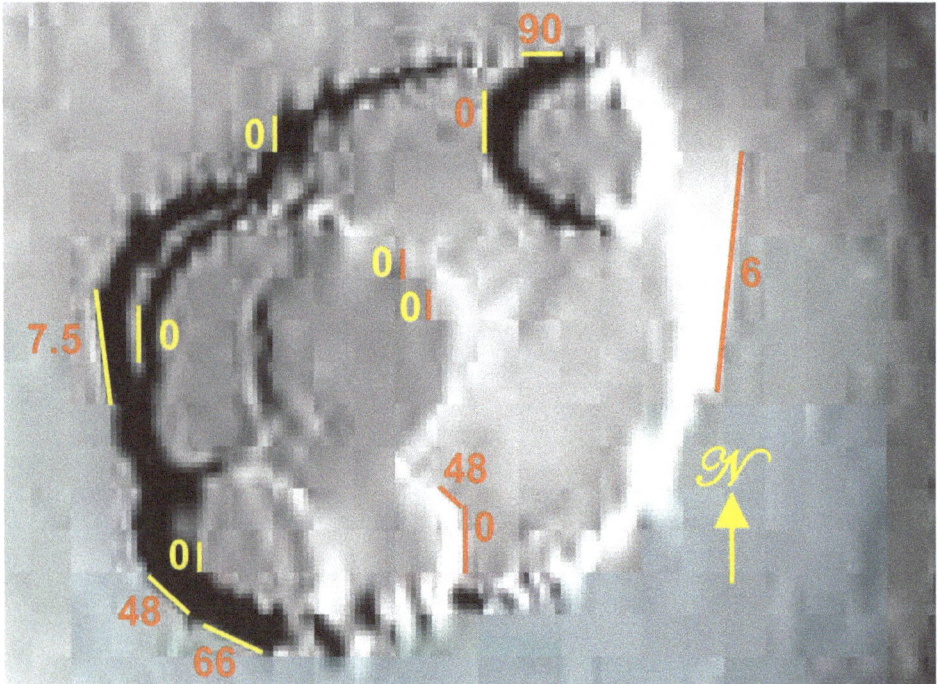

Fig. 11.9: *Straight line segments appearing in the perimeters and interiors of the calderas of Olympus Mons. The most frequent bearing angle is 0°. USGS Astrogeology.*

rectangle since their internal angles are 90°, and 45° is one-half of 90°.

But what about the other mountains? Was Olympus Mons simply an exception to the rule that mountains are supposed to be constructed out of lava flowing freely under gravitational forces from an erupting core? A closer look at the rest of the major mountains uncovered many more unexpected architectural anomalies.

Ascraeus Mons

Ascraeus Mons, like Olympus Mons, has straight line segments in its perimeter and elsewhere, but these occur less frequently than for the larger mountain. In Fig. 11.10a, a straight line edge of a groove at the boundary of the northeastern perimeter was found to have a bearing angle of 36° in the counterclockwise direction. Just south of the middle of the mountain, there is a linear section of the eastern perimeter with a bearing

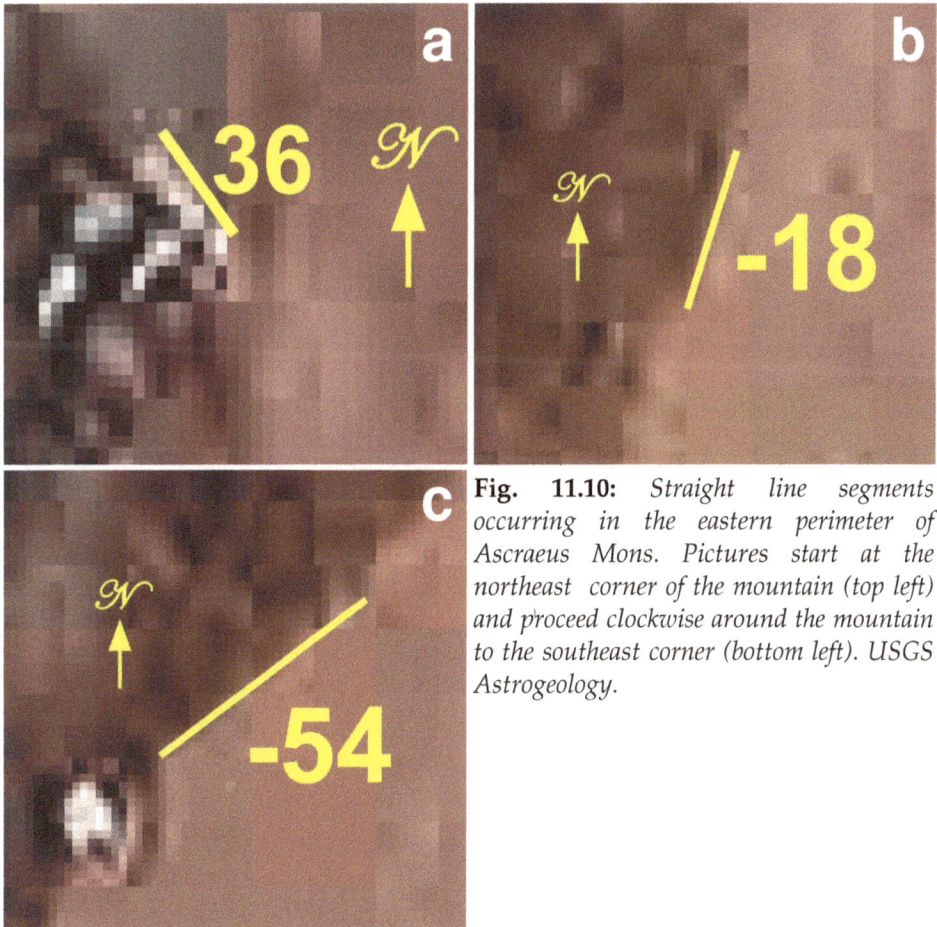

Fig. 11.10: *Straight line segments occurring in the eastern perimeter of Ascraeus Mons. Pictures start at the northeast corner of the mountain (top left) and proceed clockwise around the mountain to the southeast corner (bottom left). USGS Astrogeology.*

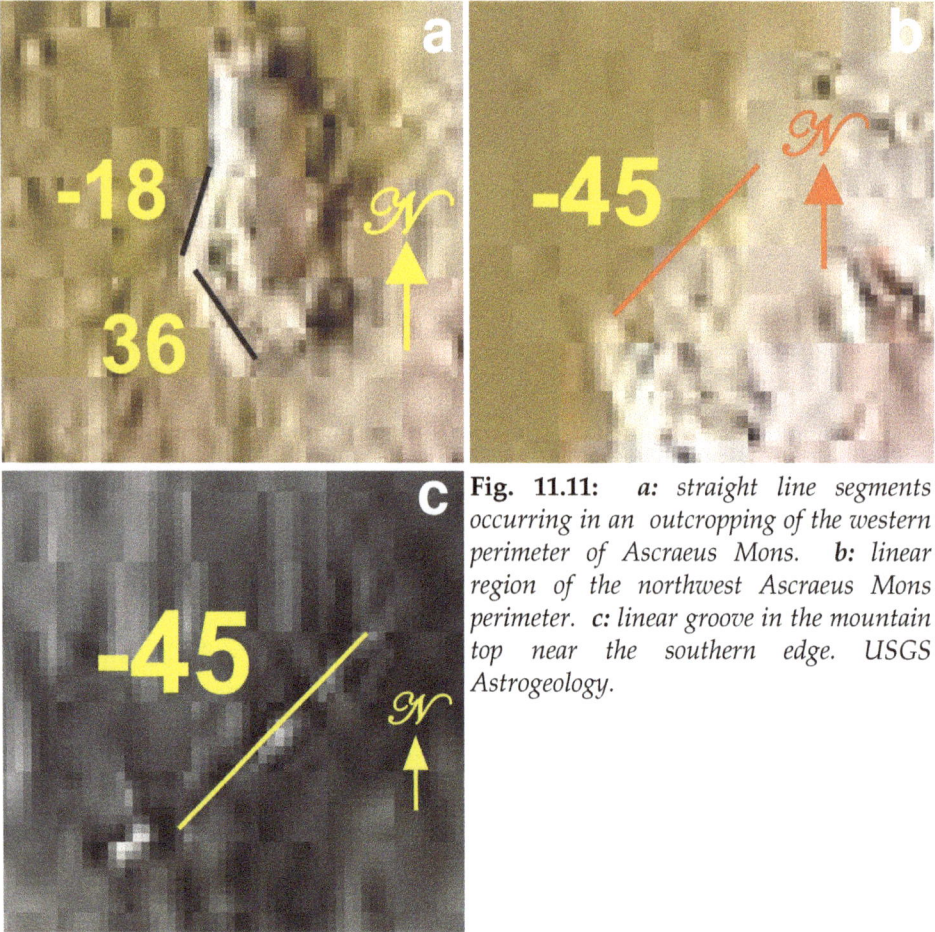

Fig. 11.11:   a: *straight line segments occurring in an outcropping of the western perimeter of Ascraeus Mons.* b: *linear region of the northwest Ascraeus Mons perimeter.* c: *linear groove in the mountain top near the southern edge. USGS Astrogeology.*

angle of 18° in the clockwise direction (Fig. 11.10b). Another linear section of the perimeter occurs at the southeast side of the mountain, and it has a clockwise bearing angle of 54° (Fig. 11.10c). In Fig. 11.11a, straight lines are shown in what appears to be an outcropping from the western perimeter rather than the actual perimeter itself. The bearing angles of these lines are 18° in the clockwise direction and 36° in the counterclockwise direction. Fig. 11.11b shows a straight section of the northern part of the western perimeter of Ascraeus Mons. It has a bearing angle of 45° in the clockwise direction. A straight line groove was also found in the mountain top near the southern border of the main edifice of the mountain with a bearing angle of 45° in the clockwise direction. All of the bearing angles found so far with Ascraeus Mons appear to relate either to the pentagram or to the square/rectangle. The value of 36° is the size of the angle of a star point in the pentagram and 18° is one-half this value. The value of 54° is one-half the size of the angle between the star points or of the interior angles of the

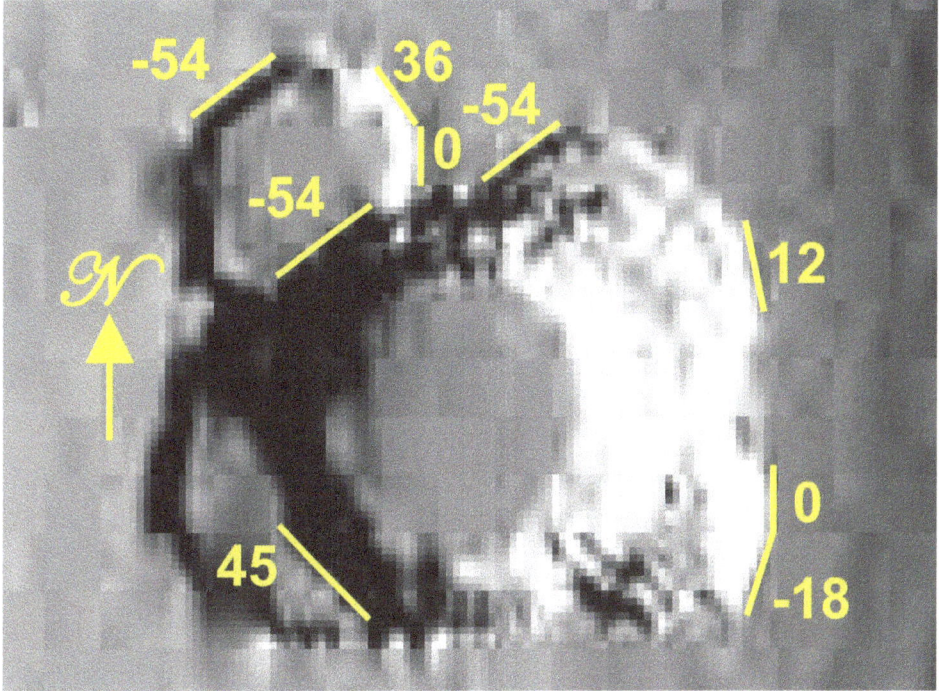

Fig. 11.12: *Straight line segments found in the perimeter and interior of the Ascraeus Mons caldera complex. Most of the values of the bearing angles are related to the pentagram. USGS Astrogeology.*

pentagon inside the pentagram. The angles of 45° may relate to the square or rectangle since 45° is one-half the value of their 90° angles.

The theme of the pentagram is carried over into the caldera complex of Ascraeus Mons (Fig. 11.12) where straight line segments in the perimeter and interior are related to the angles of the pentagram (-18°, 36° and -54°). The angle of 12° might also be related to the pentagram since it divides evenly into angles associated with that geometric shape, namely, 36°, 72° and 108°. In addition, there is an angle of 45° as was found in other parts of the mountain. The bearing angle of 0° acts like a pointer to the North Pole and South Pole.

Pavonis Mons

There are 2 straight line sections in the perimeter of Pavonis Mons which I could easily identify. They occur in the southeast section of the mountain perimeter (Fig. 11.13). The longer one travels for over 95 km so they are not trivial in size. Both angles are in the clockwise direction and are associated with the pentagram since 36° is the angle of the star points and 72° is the size of the base angles of a pentagram star point.

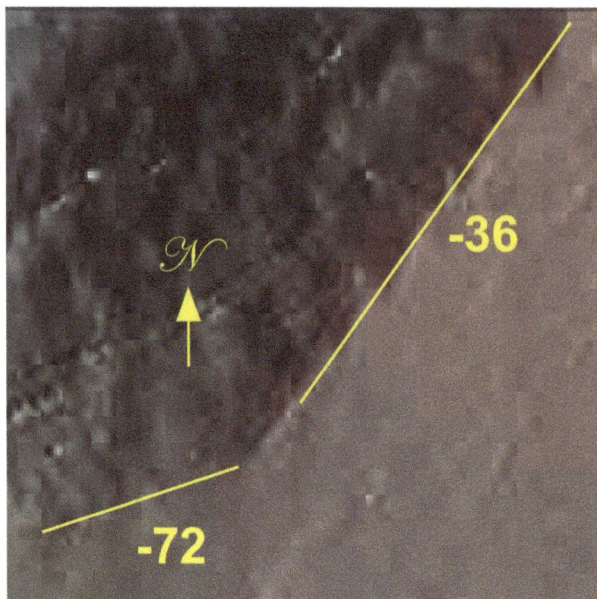

Fig. 11.13: *Straight line segments in the southeast perimeter of Pavonis Mons. The line with a bearing angle of 36° in the clockwise direction is over 95 km in length. USGS Astrogeology.*

The inner caldera of Pavonis Mons is well formed and a short length of its perimeter is linear with a bearing angle of 36° in the counterclockwise direction (Fig. 11.14). There is also what appears to be the remnants of an older caldera which extends well beyond the inner caldera. It has a straight line segment in its perimeter with a bearing angle of 54°. Both 36 and 54 are degree sizes associated with the pentagram. The outside edge of the larger caldera has a linear section with a bearing angle of 24°. This angle is not associated with the pentagram but it is evenly divisible by 3

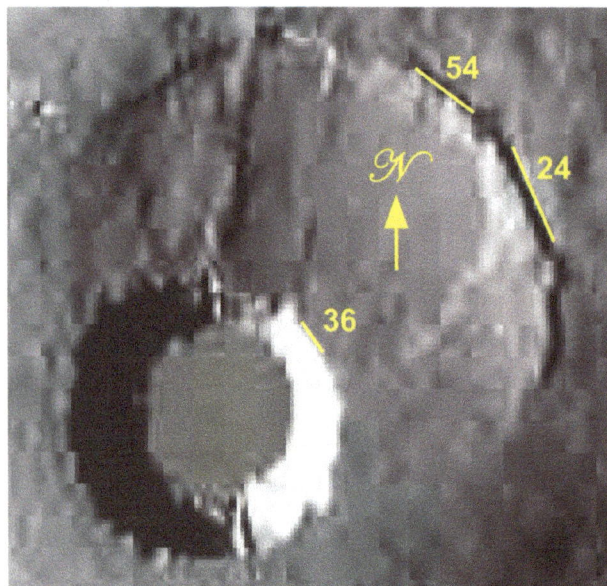

Fig. 11.14: *Straight line segments on the inner and outer caldera perimeters of Pavonis Mons. The line with a bearing angle of 24° follows the outside edge of the outer caldera perimeter which appears to be a ridge in this section. Note the rectangular area which extends from the northeast inner caldera wall to the edge of the outer caldera. USGS Astrogeology.*

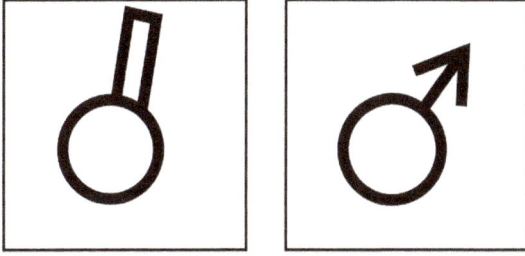

Fig. 11.15: *Left: symbol created by the Pavonis Caldera and the rectangular area to the northeast. Right: astrological symbol for Mars.*

and 6 as are the other straight line segments discussed so far.

In Fig. 11.14 there is a rectangular area extending from the perimeter of the inner caldera to the perimeter of the outer caldera. Its long sides have bearing angles of about -9° and their lengths are about π times the width of the rectangle. The entire rectangle is displaced just east of the inner caldera north-south midline. Together with the inner caldera, the rectangle creates a shape reminiscent of the astrological symbol for Mars, namely a circle with an arrow projecting outwards from its circumference (Fig. 11.15). The only thing missing is the arrowhead. Could this be an extremely ancient symbol for Mars, being only slightly different than the one now used? The shaft comes out from the circle in the same quadrant as is used in the present day symbol. I leave this for the reader to speculate about.

There are also several linear groove lines on the surface of Pavonis Mons. These will be analysed in the third book of this series.

Arsia Mons

I found only 2 sections of the perimeter of Arsia Mons to which I could fit straight lines with confidence (Fig. 11.16). The one on the southwest perimeter has a bearing angle of 45° and runs for about 115 km. The other

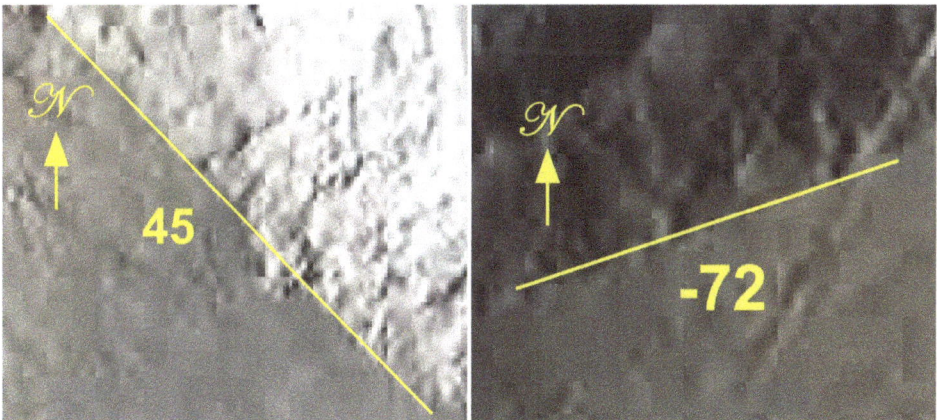

Fig. 11.16: *Left: linear southwest edge of main edifice of Arsia Mons runs continuously for about 115 km except for a small overflow area just southeast of its centre. Right: linear part of the southeast perimeter runs about 65 km. USGS Astrogeology.*

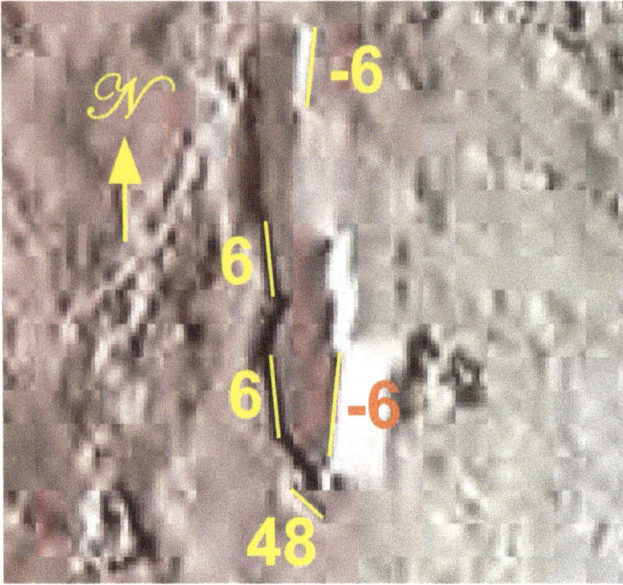

Fig. 11.17: *Straight line edges located on the top and bottom edges of cliffs surrounding a depressed region on the northwest side of Arsia Mons. All of the edges that I could measure have bearing angles of 6° or -6°. A short ridge with a bearing angle of 48° lies to the south. USGS Astrogeology.*

on the southeast perimeter has a bearing angle of -72° and runs for about 65 km. On the northwest side of the mountain, there is a relatively large depressed region which has linear cliffs forming its boundaries (Fig. 11.17). The bearing angle of either the top or bottom edge of these linear cliffs is either 6° in the clockwise direction or 6° in the counterclockwise direction. Just south of this region there is a short linear ridge with a bearing angle of 48° in the counterclockwise direction.

Short linear segments of the caldera perimeter were found, one with a bearing angle of 27° and the other with a bearing angle of 54° (Fig. 11.18). One short section of the bottom of the wall of the caldera has a bearing

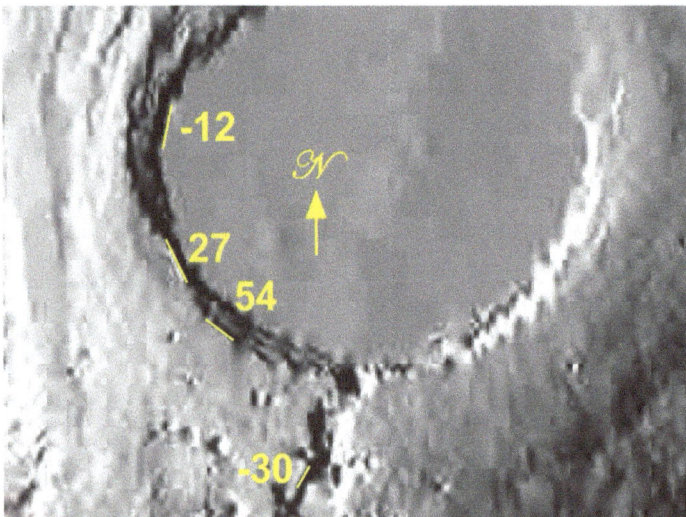

Fig. 11.18: *Straight line edges are found in the perimeter and inside edge of the wall of the Arsia Mons Caldera. A straight line segment with a bearing angle of -30° is also found in a channel running south from the caldera. USGS Astrogeology.*

angle of -12°. A channel runs south from the caldera. A short linear section of its wall has a bearing angle of -30°.

In summary, all straight lines associated with Arsia Mons have bearing angles evenly divisible by 3, with many of them being divisible by 6 as well. A much smaller proportion of these bearing angles are associated with the pentagram than was found for Pavonis Mons.

Biblis and Ulysses Tholi

Olympus Mons and the Tharsis Montes are not the only mountains to have straight line segments in their makeup. The smaller mountains also demonstrate this feature. To illustrate this I will start with Biblis and Ulysses Tholi, a pair of mountains north of Arsia Mons and west of Pavonis Mons. The upper surface of the west side of Biblis Tholus contains several trenches whose sides have linear sections either along their top or their bottom edges (Fig. 11.19). I found 5 straight line segments here with bearing angles of 24° or 36°, both divisible by 12, 6, 4, 3 and 2.

The edge of the east side of Biblis Tholus (Fig. 11.20) is constructed from 3 long straight lines. The northern line matches the southern line in bearing angle (24°) except it is in the counterclockwise direction whereas the southern line is in the clockwise direction. The middle line has a bearing angle of 0° and therefore can act as a pointer to the poles. There

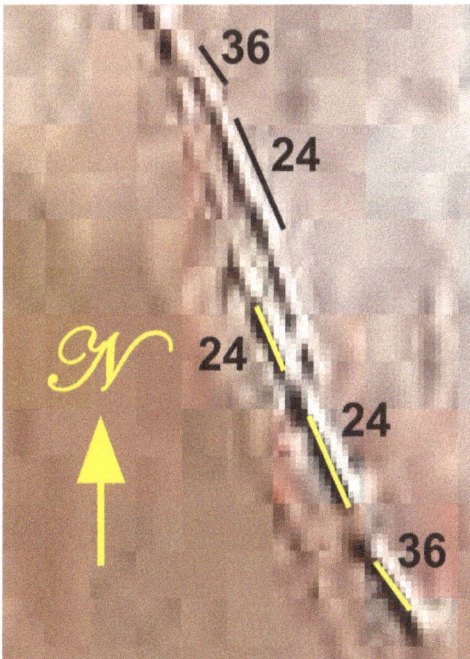

Fig. 11.19: *The west side of Biblis Tholus is marked by long trenches whose top or bottom edges form straight lines. These lines were found to have bearing angles of 24° or 36° in the counterclockwise direction. Both angle sizes are evenly divisible by 12, 6, 4, 3 and 2. USGS Astrogeology.*

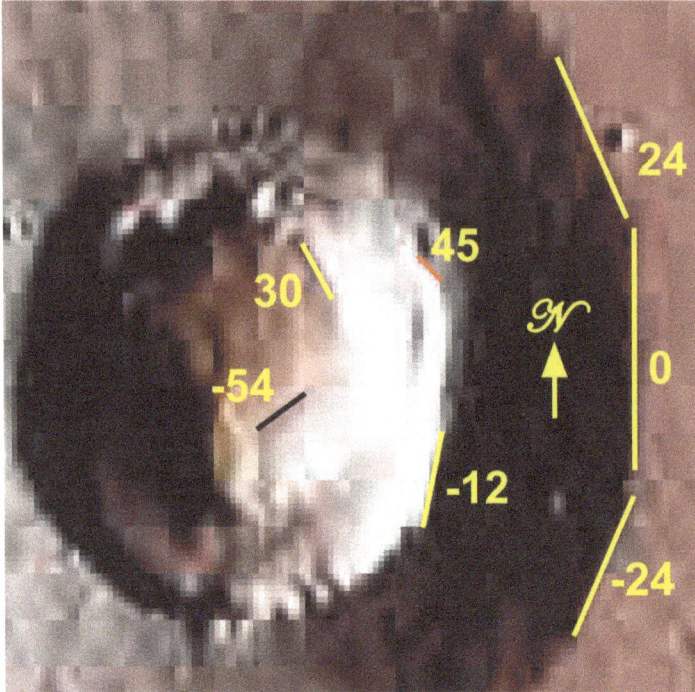

Fig. 11.20: *The perimeter of the east side of Biblis Tholus is constructed from 3 long straight lines. The perimeter of the caldera has 2 short linear sections. Inside the caldera are 2 linear structures. All of the nonzero bearing angles are evenly divisible by 3. USGS Astrogeology.*

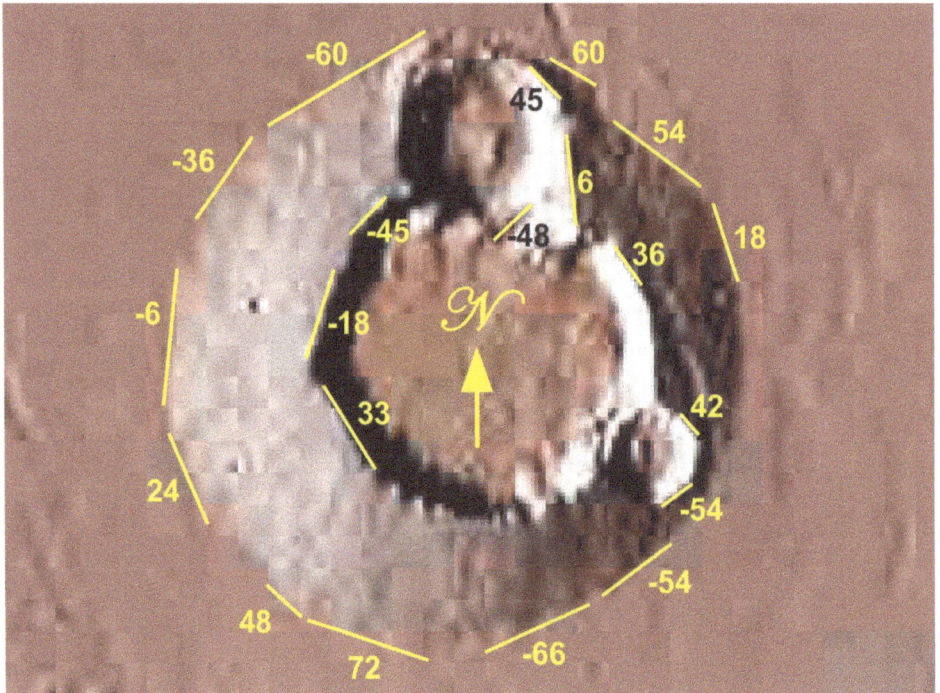

Fig. 11.21: *The perimeter of the Ulysses Tholus is constructed mainly from straight line segments. Straight lines are also present in the perimeters of the caldera and surface craters. USGS Astrogeology.*

are 2 short linear sections of the caldera perimeter with bearing angles of -12° and 45°, and 2 linear structures inside the caldera with bearing angles of 30° and -54°. Except for 0°, the bearing angles of all these straight lines are evenly divisible by 3 and most by 6 as well.

When we get to Ulysses Tholus (Fig. 11.21) we see an extraordinary thing. The perimeter of the mountain is not curved as you would expect but rather it is constructed almost entirely out of straight line sections. The absolute values of the bearing angles of these lines range from 6° to 72° and form a series of numbers going up by 6, missing only 12, 30 and 42. The number 42 is found in the bearing angle of one of the straight line sections of the Ulysses Southeast Crater perimeter, and a repeat of the number 54 occurs here as well. A large part of the perimeter of the Ulysses Caldera is also constructed from straight line sections with bearing angles of -18°, 33°, 36° and -45°. The perimeter of the Ulysses Tholus North Crater has straight line sections with bearing angles of 6° and 45°. There is also a linear depression occurring between the northern crater and the caldera which has a bearing angle of 48° in the clockwise direction.

Uranius Mons and Ceraunius Tholus

Examination of the cluster of 3 mountains to the northeast of Ascraeus Mons revealed a single short linear section in the northeast section of the perimeter of the Uranius Mons Caldera with a bearing angle of 45° (Fig. 11.22). I could not reliably measure any other straight line sections either for this mountain nor for Uranius Tholus. However, the perimeter of Ceraunius Tholus has 4 sizeable straight line segments, 1 in the southwest perimeter with a bearing angle of 54°, and 3 in the eastern perimeter

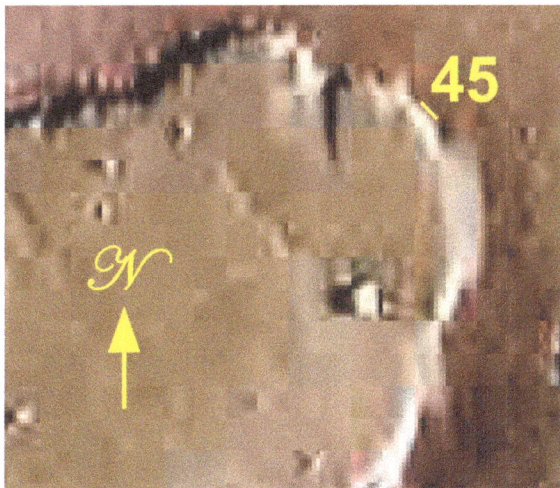

Fig. 11.22: *The caldera of Uranius Mons has a short straight line section with a bearing angle of 45° in the counterclockwise direction. No other straight lines could be reliably measured for this mountain. USGS Astrogeology.*

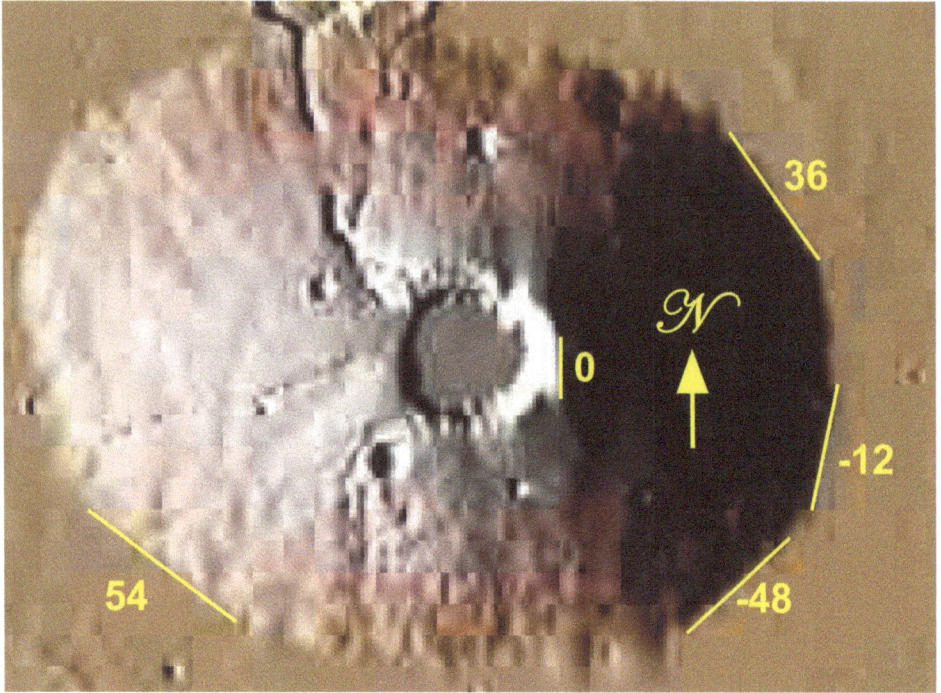

Fig. 11.23: *Ceraunius Tholus has 4 straight line segments in its perimeter, all with bearing angles divisible by 6. The east side of the caldera has a straight line with a bearing angle of 0°. USGS Astrogeology.*

which form a group of numbers divisible by 12 (Fig. 11.23). Most of the east side of the Ceraunius Tholus Caldera is linear with a bearing angle of 0°, so it could act as a pointer to the poles.

Tharsis Tholus

Tharsis Tholus presents a definite theme with the bearing angles of its straight line segments (Fig. 11.24). I was able to measure the bearing angles of 2 straight line sections of its western perimeter. Both were 36° in the counterclockwise direction. The northern perimeter shows an intermittent straight line with a bearing angle of -60°. Just west of the north part of the western perimeter is a ridge with a bearing angle of 30°. Straight lines were found in the perimeters of 2 craters on the surface of the mountain, one with a bearing angle of -54° and the other with bearing angles of 0° and 36°. The edge of a large section of the west side of the mountain which appears to have fallen away has a straight line section with a bearing angle of -36°. Yet another bearing angle of -36° is to be found inside the caldera. This is found with the bottom linear edge of a

Fig. 11.24: *The bearing angles of the straight lines found on Tharsis Tholus are mainly associated with the pentagram. The bearing angles of 30° and -60° could be associated with the equilateral triangle. USGS Astrogeology.*

sizeable structure in the southern part of the caldera.

In total, 5 out of the 9 straight lines associated with Tharsis Tholus have bearing angles with a magnitude of 36 degrees, the size of the angle of a star point in the pentagram. Another of the straight lines has a bearing angle of 54° which is half the angle between the star points in a pentagram. Thus 2/3 of the lines have bearing angles associated with the pentagram. The 2 lines with bearing angles of 30° and 60° are probably associated with the equilateral triangle where 60° is the size of all the internal angles and 30° is one-half of 60°.

Apollinaris Mons

A long landform created by lava flow from Apollinaris Mons lies to the north of the mountain. The western edge of this lava mass has linear sections with bearing angles of -9, 12 and -18 degrees (Fig. 11.25 left). The longest section stretches for about 45 km. The west bottom edge of the Apollinaris Caldera has a bearing angle of about 7.5° in the clockwise direction (Fig. 11.25 right). The number 7.5 came up once before for the bearing angle of the west side of the Olympus Mons Caldera complex. Interestingly, these 2 instances are both for the west sides of calderas, However, the straight line for Olympus Mons is for the top edge of the

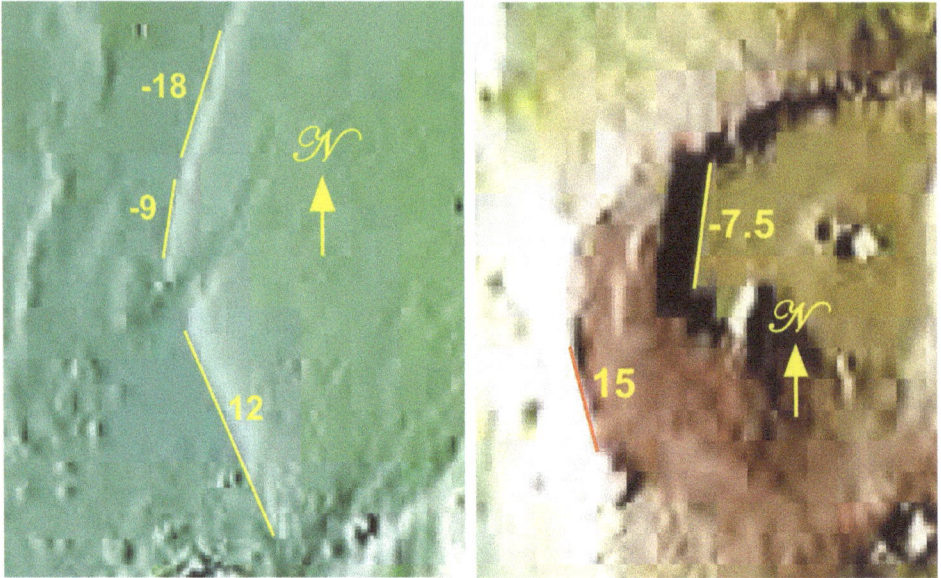

Fig. 11.25: *Left: the northern lava flow from Apollinaris Mons has straight line edges with bearing angles divisible by 3. Right: The west edge of the top of the edifice has a bearing angle of 15°, the magnitude of which is double that found for the bottom edge of the caldera west wall. USGS Astrogeology.*

caldera wall and is in a counterclockwise direction. Also interesting is the 15° bearing angle of a straight line at the west edge of the top of the mountain edifice. The angle magnitude is double that of the caldera straight line and the linear region also on the west side. The number 7.5 will occupy a prominent place in a chapter of the third book of this series.

The Elysium Group of Mountains

The group of mountains consisting of Elysium Mons, Albor Tholus and Hecates Tholus has a few straight line segments as well. Starting with Hecates Tholus, a straight line with a clockwise bearing angle of 45° is seen in a small crater south of its caldera (Fig. 11.26). Albor Tholus has a straight line with a bearing angle of 36° occurring on the side of a cliff on the northeast side of the mountain (Fig. 11.27). A short linear section with a bearing angle of 45° in the clockwise direction is found in the northwest corner of its caldera perimeter. A similar short linear section with a bearing angle of 45° in the counterclockwise direction occurs in the Elysium Mons Caldera perimeter (Fig. 11.28), but no linear sections were found either in the perimeter of that mountain or elsewhere. It is interesting to note that straight lines with a bearing angle magnitude of 45° were found on all 3 mountains of this group.

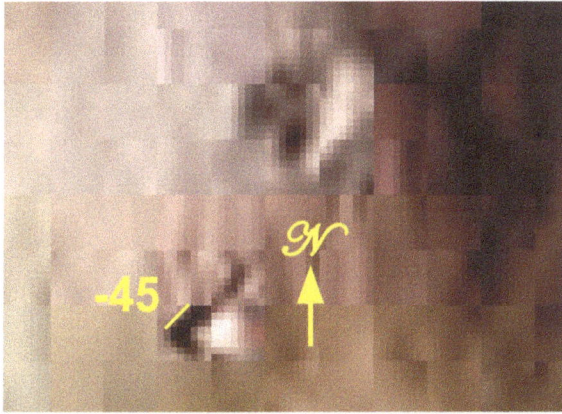

Fig. 11.26: *A small crater south of the caldera of Hecates Tholus has a straight line section in its northwest perimeter. The bearing angle is 45° in the clockwise direction. USGS Astrogeology.*

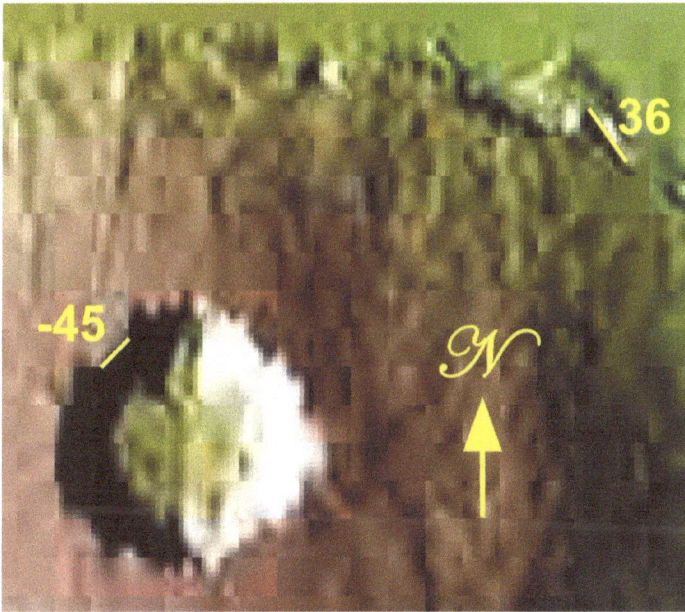

Fig. 11.27: *There is a straight line with a bearing angle of 36° that runs on the side of a cliff on the northeast side of Albor Tholus. A short linear section of the caldera perimeter is present on its northwest side and has a bearing angle of -45°. USGS Astrogeology.*

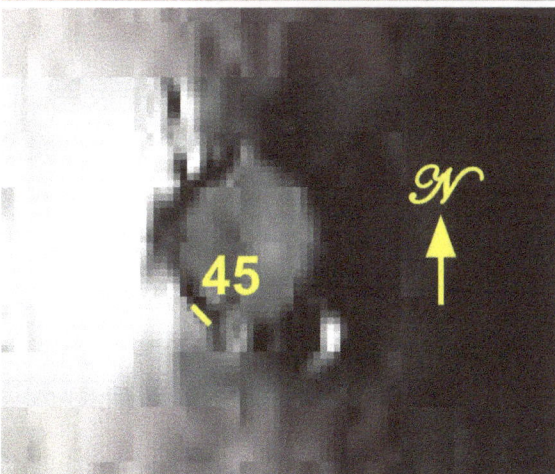

Fig. 11.28: *The perimeter of the Elysium Mons Caldera contains a very short linear section with a bearing angle of 45° in the counterclockwise direction. USGS Astrogeology.*

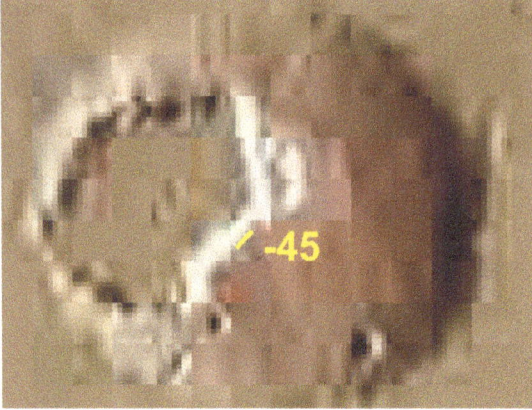

Fig. 11.29: *The Jovis Tholus Caldera has a short straight line section in its southeast perimeter. Its bearing angle is 45° in the clockwise direction. USGS Astrogeology.*

Jovis Tholus

The smallest mountain on the Tharsis Rise is Jovis Tholus. Its caldera perimeter has a straight line section on the southeast side with a bearing angle of -45° (Fig. 11.29). This size of bearing angle is found in all the mountains except for Pavonis Mons, Apollinaris Mons and Tharsis Tholus.

Summarizing the Data

A total of 140 straight line structures were found in the major Martian mountains. When the frequency of occurrence of the absolute value (i.e., the magnitude without regard for sign) of their various bearing angles is tabulated, interesting patterns start to emerge (Table 11.1). The most obvious one is that all non-zero bearing angles are evenly divisible by 3 except for the angle of 7.5°. Most of the non-zero angles are also divisible by 6. For those that are not, other than 7.5° and 45°, there is only a single occurrence. For numbers divisible by 6, there is a minimum of 1 occurrence (for 90°) and a maximum of 20 occurrences (for 36°) but most of these angles occurred 3 to 13 times. Hence, there seems to be a focus on numbers divisible by 6. There also seems to be a focus on the pentagram since the maximum number of events were for 36°, the size of the angle of the star points of the pentagram. Other numbers associated with the pentagram (18, 54, and 72) were also well represented. The values of 0° and 45° had the next highest rates of occurrence with 17 instances for 0° and 18 instances for 45°. The latter angle could represent a square or rectangle since 45° is one-half of a 90° angle. The values of 30° and 60° occurred 7 times each. They can be associated with the equilateral triangle since its internal angles are 60°, and 30° is one-half that value. That leaves 6°, 12°, 24° and 48° to account for. All of these numbers except 6 are multiples of 12

Table 11.1: *Frequency of occurrence of bearing angles for straight line structures in the mountains. Clockwise and counterclockwise values are grouped together.*

Bearing Angle (°)	# of events	Bearing Angle (°)	# of events	Bearing Angle (°)	# of events
0	17	24	9	51	1
3	1	27	1	54	13
6	9	30	7	60	7
7.5	2	33	1	66	2
9	1	36	20	72	4
12	8	42	3	75	1
15	1	45	18	90	1
18	7	48	6		

which is a very sacred number in its own right. The number 12 will be discussed in detail in Chapter 12 which deals with the Martian meter. The number 6 is half the value of 12. It is likely, therefore, to also be associated with 12 as there seems to be a tendency by the hypothetical architects on Mars to emphasize half values of important numbers such as 36 and 108.

The high number of bearing angles with a value of 0° suggests that this bearing angle was given quite a bit of importance by the postulated architects. A likely possibility for this could be that it was used as a pointer to the poles. Large numbers of such pointers would give overhead spacecraft plenty of indicators to help orient themselves with regard to north and south.

Implications of Finding Straight Line Segments

The multiple occurrence of specific bearing angles for the straight line segments found on the mountains is astronomically improbable under random conditions. This will be discussed in more detail in Chapter 13. Hence, the most logical conclusion is that these mountains have not only been artificially placed but must also have been artificially constructed! Many of the line segments associated with these Martian mountains, especially with Olympus Mons, mark out features which are several kilometers high. Hence, they cannot simply be the result of ordinary construction techniques which might be employed in erecting pyramids or buildings. Rather, the construction technology had to have been far in advance of what the human engineers of today could achieve. It would appear that the Martian engineers were able to melt and shape solid rock in huge amounts at a time. If you look carefully at the top surface of

Fig. 11.30: *Upper surface of Olympus Mons appears to have been constructed in roughly concentric layers. For this to have happened naturally, each succeeding eruption would have had to have been systematically less powerful and possibly the ejected material had to have been more viscous than its predecessor. USGS Astrogeology.*

Olympus Mons, you will see that it is actually terraced rather than covered with continuous lava flows (see Fig. 11.30). The entire upper surface seems to have been constructed out of enormous concentric irregularly-shaped polygonal layers, with each subsequent layer covering a smaller area than the lower layer immediately beneath it. There are up to 8 distinguishable layers on the east side of the mountain, and as many as 7 layers on the west side. A similar type of construction appears to have been used for the 3 Tharsis Montes as well.

The issue of the source of the vast quantities of material which the mountains are constructed out of must also be addressed. Since the calderas themselves have linear segments with standardized bearing angles, they must have been artificially constructed as well. This calls into question whether or not the substance of the mountains is from magma beneath the mountains or whether it has been quarried from another area of Mars and transported to the mountain sites. Although it is possible that magma may have been channelled up through a central tube simulating the process of a natural volcanic eruption, this option would require some type of pressure to drive the magma up through the tube to great heights.

Such pressure could perhaps come from the radioactive heating of the planetary interior. However, it is possible that the material for mountain construction could have been excavated from a site such as the northern "ocean" regions, transported and melted down and then sculpted to form the giant mountains. The Hellas and Argyre impact basins as well as artificially constructed craters are other potential sources of material. Then there are the vast Valles Marineris canyons which could have contributed up to 3 million cubic km of material[1]. Alternatively the material might have originated from beneath the Martian crust, e.g., to carve out a vast biosphere. Specimens will ultimately have to be taken from the mountains to be examined and compared to samples derived from other sites on Mars to scientifically establish the true origin of the substance that the mountains are made of.

If we accept the seemingly outlandish possibility that the mountains were constructed rather than the result of volcanic eruptions, then the possibility also exists that a vast living space could have been engineered within them. This could be in the form of caves or in the form of gigantic condominiums with integrated shopping, recreational and green space areas, the likes of which are well beyond the wildest dreams of terrestrial developers. This is more fully discussed in Chapter 15.

References

1. *Schultz, R.A. 1991. "Structural Development of Coprates Chasma and Western Ophir Planum, Valles Marineris Rift, Mars". Journal of Geophysical Research 96: 22,777-22,792.*

12

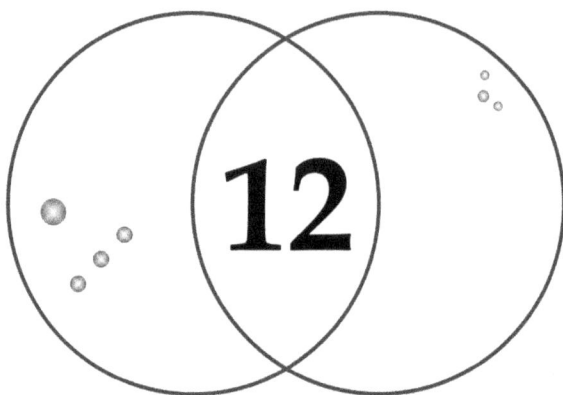

The Martian Meter

The sacred geometry patterns seen with the mountains of Mars can only be viewed from outer space. Hence, their creators must have been capable of space travel and would very likely have visited the planet Earth as well as Mars. If so, it is possible that they would have imported some of their technology such as a standard unit of measurement which was used on this planet by past civilizations having a connection to Mars. As a long shot I decided to investigate whether the Megalithic yard would make any sense as a unit of measurement on Mars. The Megalithic yard was determined by Alexander Thom to be the unit of measurement used by the architects of the megalithic sites, including Stonehenge, throughout the British Isles and even in Brittany, France. It has a precise value of 2.720 feet with a standard error of ± 0.003 feet[1]. I wanted to see if it had a relationship to the equatorial radius of Mars since I reasoned that an advanced race might base their standard unit of measurement on an important dimension of their home planet. I therefore divided the currently accepted value of the equatorial radius of Mars, 3396.19 km, by 0.8291 meters, the metric equivalent of 2.720 feet. The answer worked out to 4096.24×10^3 Megalithic yards (or 4096.24 Megalithic kiloyards, using the prefix 'kilo' to represent 1000). As this was not an even number like 4000 or 4200, I was prepared to walk away from this wild idea when it suddenly struck me that the number I got was extremely close to 4096, and exactly 4096 when you restrict the answer to only 4 significant figures which was the number of significant figures contained in Thom's measurement. This brought me back to my days in graduate school when I worked on a Digital Equipment PDP-8i computer which operated on 12 bit words instead of the 32 or 64 bit words of modern PCs. The 12 bit word was capable of having 4096 different configurations since $2^{12} = 4096$.

The number 12 is one of the only 2 numbers known to be sublime. A sublime number is one which is a positive integer having a perfect number of divisors which themselves add up to a perfect number. To understand this, we first have to know what a perfect number is. A perfect number is a positive integer which is the sum of all its positive divisors except for the number itself. The number 12 has 6 divisors including itself. The number 6 is a perfect number since it has 3 divisors other than itself which add up to itself, i.e., $1+2+3 = 6$. The 6 divisors of 12 add up to 28 (i.e., $1+2+3+4+6+12 = 28$) which is also a perfect number (i.e., $1+2+4+7+14 = 28$). The only other sublime number which we know of has 76 digits. There are many other mathematical properties of the number 12 which are too complex to present here.

In our own history, 12 has been used to represent governmental perfection, both in religious and secular structures. Thus we have the 12 tribes of Israel, the 12 apostles, and the 12 knights of King Arthur's round table. We also number the hours in a half day from 1 - 12. It is important to note that in astrology the number 12 is the number of signs in the zodiac which divide the ecliptic (the plane of the Earth's orbit) into 12 approximately equal zones of celestial longitude, each represented by a constellation of stars. Whether or not the Martians divided the heavens into 12 constellations 3 - 4 billion years ago as we do today can only be speculated at this point. At any rate, constellations of some sort would have been around 3 - 4 billion years ago and would be the same on Mars as seen from Earth. However it would take about 14,600 Earth years to complete a term for any particular zodiac constellation on Mars instead of the ~2,160 years it takes here since the full precessional cycle takes about 175,000 Earth years on Mars instead of 25,920 years on Earth. During the solar year on Earth the sun rises in a different constellation at about the 22nd day of each month, and this could be the origin of the practice of dividing the year into 12 months. If this was done on Mars, the zodiac months would each last about 55.75 Martian days (sols), or about 57.25 earth days. Since the number of sols is not an integer they would have needed to construct months of unequal days as we have now.

The finding that the equatorial radius of Mars can be expressed as an binary integer of Megalithic kiloyards has profound implications since it would link the megalithic builders on earth to engineers from the planet

Mars. And if the Megalithic yard was a measure brought from outside the solar system, it opens the possibility that the planet Mars itself was an engineered planet. However, at the moment, it appears that the more likely possibility is that it was derived from the planet's radius in much the same way that the meter was intended to represent 10^{-7} of the earth's quadrant (one quarter of the earth's circumference) passing through Paris.

On the basis of the above findings, I propose that a new unit be established that would be called the Martian meter to replace the Megalithic yard. It would have the value of (3396.19 km x 1000 m/km)/(2^{12} x 1000) which works out to 0.829148 meters. This value is then accurate to the 6 significant figures of the currently accepted value for the equatorial radius of Mars[2] instead of the 4 significant figures for the Megalithic yard. In addition it is so close to the value of Thom's Megalithic yard (differs by only 0.0003 feet or 0.0036 inches) that the dimensions of megalithic sites in terms of Martian meters will resolve to the same integer values or simple fractions of integers that Thom found. It is a tribute to Thom's expertise and diligence that he was able to determine the Megalithic yard to such an amazing accuracy. I preferred to denote this unit as a meter rather than as a yard to avoid confusion with the Megalithic yard, especially with respect to an abbreviation. Thom used the abbreviation "MY" for the megalithic yard. I suggest that "Mm" be used to abbreviate the Martian meter since it would permit the different powers of 10 of Martian meters to have the same abbreviations that we use for the terrestrial meter e.g., Mmm for Martian millimeter, Mcm for Martian centimeter, Mkm for Martian kilometer.

The many sacred distances that are a function of the equatorial radius R can easily be expressed in terms of Martian kilometers. Thus:

$$R/2 = 2^{11} \text{ Mkm}$$
$$R/\sqrt{5} = 2^{12}/\sqrt{5} \text{ Mkm}$$
$$R/(2\sqrt{5}) = 2^{11}/\sqrt{5} \text{ Mkm}$$
$$R/(3\pi) = 2^{12}/(3\pi) \text{ Mkm}$$
$$R/(4\pi) = 2^{10}/\pi \text{ Mkm}$$
$$R/(3e) = 2^{12}/(3e) \text{ Mkm}$$
$$R/(5\varphi) = 2^{12}/(5\varphi) \text{ Mkm}$$
$$R/10 = 2^{11}/5 \text{ Mkm}$$
$$2R/15 = 2^{13}/15 \text{ Mkm}$$
$$\text{etc.}$$

If the fractional part of these values were worked out in terms of binary fractions, the numbers would be totally binary in representation. The fact that the radius of Mars can be expressed almost exactly as a binary integer of Megalithic kiloyards opens up the possibility that a binary number system was used by this civilization, perhaps in addition to other bases such as 10 or 60. The advent of the computer has brought the binary, octal and hexadecimal systems to our own civilization in addition to the decimal system, so this would not be preposterous speculation.

I next tried to determine if the Martian kilometer had any association with the Martian moons, Phobos and Deimos. What I found was that their distance from the centre of the planet could be expressed as an even number of Martian km multiplied by the square root of 2. Thus Phobos is almost exactly √2 x 8,000 Mkm, and Deimos √2 x 20,000 Mkm, from the centre of Mars. The measured values are not more than five one-hundredths of one percent different from these theoretical values! This suggests that these 2 moons have been artificially placed into their orbits and that the Martian meter is indeed a standard unit of measure that has been used in this solar system, both on Mars and on Earth.

When we can obtain accurate measurements on the numerous artifacts which seem to be present on Mars (e.g., the D&M Pyramid, see Chapter 15), we will be able to determine if any have been sized in integer units of the Martian meter. So far, we have only crude estimates of size since they are based on satellite data which are at too low a resolution for this purpose. I would be very surprised not to find evidence of its use on Mars when we have all sorts of evidence of its use here on planet Earth. It would only make sense that a culture which appears to be so committed to the use of sacred geometry in site placement would use a unit of measurement that relates to the planetary radius in the dimensions of the sites themselves and of other smaller structures. Only better data which should be available in the future can settle this issue.

References

1. *Megalithic Sites in Britain. Alexander Thom. Oxford University Press, 1979.*

2. *Seidelmann, P.K. (chair), Abalakin, V.K., Bursa, M., Davies, M.E., De Bergh, C., Lieske, J.H., Oberst, J., Simon, J.L., Standish, E.M., Stooke, P., and Thomas, P.C., 2002, Report of the IAU/IAG Working Group on Cartographic Coordinates and Rotational Elements of the Planets and Satellites—2000: Celestial Mechanics and Dynamical Astronomy, v. 82, p. 83–110.*

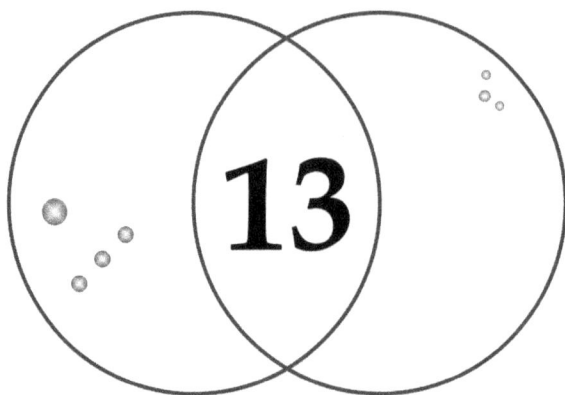

Numbers and Statistics

The human brain has no problem in recognizing that the mountain and crater patterns which I have just presented are of artificial origin. That's because our brains are pre-eminent pattern recognizers, and we know from experience that patterns like this only occur very rarely under random conditions. We would never expect to find a multitude of them in the same general locale. However, it still would be reassuring if the artificiality of what we have seen could be established by objective statistical tests. To meet this need, I have worked out several approaches to test for the artificiality of the actual placement of major sites, and have used standard equations to ascertain the probability of such arrangements occurring simply by chance. Each of these tests for artificiality provides useful information on the situation even though some were not strong enough to achieve scientifically accepted levels of statistical significance. Others reached extraordinarily high levels.

The question of artificiality arises not only for the patterns which have been observed in the placement of sites but also for the occurrence of integer or integer + 1/2 longitude differences, and for latitude coordinates that can be expressed as simple sacred geometry formulae. And how likely is it under random conditions that the linear sections which have been observed for mountain and caldera perimeters etc. would have bearing angles which are mainly restricted to a relatively small set of values? These are the type of questions that will be addressed in this chapter.

Testing the Incidence of Sacred Distances Between Sites

When I did this test, I had calculated the coordinate positions of only 29 major sites. These consisted of the major mountains of Mars (including Issedon Tholus), several important craters and the Pentagram Pyramid. Mathematically, these 29 sites have a total of 406 separate distances between them. The fact that many of these distances can be expressed in terms of sacred formulae (e.g., are an integer function of R alone, or of R multiplied or divided by 1 or 2 of the following 6 irrational numbers: e, π, φ, $\sqrt{2}$, $\sqrt{3}$, or $\sqrt{5}$) suggests that this might be a way of establishing artificiality. If it can be shown that the incidence of occurrence of sacred distances is greater than what could be expected under random conditions, it would indicate that more than natural forces have brought about this architecture. To do this, I first had to determine what proportion of the number line is composed of sacred formulae. I restricted myself to sacred formulae using the equatorial radius of Mars for this analysis to simplify matters. Since distances can only be measured to an accuracy of about ±1 km, the interval occupied by a sacred formula on the number line for this test should not be a single point but needs to be set to ±1 km thus covering 2 km.

So what sacred distances need to be considered? From the observations made with the 29 sites, it would appear that the following is required:

- R by itself
- R multiplied or divided by one of the 6 irrational numbers (n = 12)
- R multiplied by each of the ratios of the irrational numbers (n = 30)
- R multiplied or divided by each of the products of 2 irrational numbers (n = 30)
- R multiplied or divided by the square of π, φ or e (n = 6).

This gives a total of 78 primary sacred distances formed by multiplying and/or dividing R by 1 or 2 sacred numbers. Add 1 sacred distance from R by itself and you get a grand total of 79 primary sacred distances. Not too bad so far. However, such distances will only cover a limited portion of the entire range that needs to be considered. For the 29 sites under study, latitudes ranged from 8.10° S for Arsia Mons to 39.45° N for Alba Mons, covering a distance of about 2800 km. Longitudes ranged from 147.17° E for Elysium Mons to 273.47° E for the Fesenkov crater, covering a distance of about 7500 km at the equator. As it turned out, however, no site-to-site distance exceeded 7000 km so I used this figure as an approximation of maximum distance. Hence, each primary sacred distance needed to be divided and multiplied by a range of integers sufficient to cover the entire span of the number line up to 7000 km. This produced a huge increase in the number of sacred distances, especially for the shorter distances. I also found it appropriate, in the light of discoveries reported in the next 2 books of the *Intelligent Mars* series, to include the sacred distances resulting from multiplying R by each of the music interval ratios found in the major and minor diatonic scales (9/8, 6/5, 5/4, 4/3, 3/2, 8/5, 5/3, 9/5, and 15/8) plus a minor second (16/15), augmented fourth (45/32) and diminished fifth (64/45), as well as by the inverses of all of these ratios. This only increased the total number of sacred distances by another 24 values.

The results of this analysis are shown in Table 13.1. The 3rd column lists, for 15 ranges of distance, the proportion of the number line occupied by sacred formulae assuming each to cover a span of ±1 km. Some of the formulae overlap, so the coverage is not simply a product of the quantity of sacred formulae times 2 km. For distances greater than 4000 km, the coverage is less than 10%. However, as distances get shorter, greater and greater proportions of the number line are occupied by sacred formulae until at about 400 km, the coverage is 100%.

When I examined the distances between the 29 sites on Mars, I found that the percentage that can be fit to within 1 km of a sacred distance formula (column 6 of Table 13.1) was greater than the percentage of the number line covered by sacred formulae (column 3) for 8 of the 12 ranges which did not have 100% coverage. However, a statistical test applied to each interval showed that the number of sacred distances observed was not greater than could be expected from just random occurrence for any of these distance ranges.

Hence, although there is a slight tendency for more than expected sacred distances to be present this could not be confirmed statistically. The problem lies in the fact that in order to maintain objectivity, I could not eliminate any sacred formulae from the test even though many of them were extremely unlikely to have been used by hypothetical architects who would only want to focus on the ones which conveyed a certain message. I could not be certain what that message was, and I would be guilty of the absurdity of presuming artificiality in a test for artificiality if I discarded any on that basis. Another problem with this test is that column 5 of Table 13.1 contains 27 survey distances (e.g., the distances to Biblis Tholus and Ulysses Tholus from Arsia Mons and Pavonis Mons). These distances are biased since they were directly set to sacred formulae.

Due to the shortcomings of this analysis, I had no other choice but to devise tests that were more sensitive. But this first exercise was not a wasted effort as it provided important information on the quantity of sacred formulae as a function of distance. The test of incidence of sacred distances on Mars will be revisited in the next chapter once artificiality has been established by other tests, allowing a more workable subset of sacred distances to be selected for testing.

Table 13.1: *Percent occurrence of sacred distances between 29 sites on Mars compared to percent coverage of the number line by sacred formulae. Each sacred formula is assumed to cover 2 km on the number line. The sacred distances on Mars did not show a statistically significant higher rate of occurrence over random for any distance range.*

Interval (km) From	To	% number line used by sacred formulae	Number of site-to-site distances	Number of sacred distances (±1 km)	Percent sacred distances
0.00	199.99	100.00	7	7	100.00
200.00	299.99	100.00	5	5	100.00
300.00	399.99	100.00	7	7	100.00
400.00	499.99	98.42	7	7	100.00
500.00	599.99	97.99	14	14	100.00
600.00	699.99	90.41	15	14	93.33
700.00	799.99	81.81	17	15	88.24
800.00	999.99	68.99	28	19	67.86
1000.00	1499.99	44.25	64	27	42.19
1500.00	1999.99	28.81	54	16	29.63
2000.00	2499.99	19.90	42	7	16.67
2500.00	2999.99	15.48	15	5	33.33
3000.00	3999.99	12.47	24	2	8.33
4000.00	5499.99	9.40	56	7	12.50
5500.00	6999.99	8.30	51	6	11.76
	Total		406	158	

Testing the Incidence of Isosceles Triangles

It seems that the high incidence of isosceles triangles between sites is well beyond what you would expect to find under natural conditions. To test for this, I counted the number of isosceles triangles that were obtained from the same 29 sites under consideration in the previous tests. I limited the definition of 'isosceles' to those triangles in which the equal sides differed only by 1 km or less in length (since 1 end of each of the 2 sides was common to both, I only had to consider the accuracy error of the other ends which I assumed to be ±0.5 km each). Since a large number of triangles can be created from 29 sites it can be expected that some will be isosceles by pure chance. The question is, are there more isosceles triangles present than could be expected from pure randomness? From the 29 Martian sites, I found a total of 20 isosceles triangles in which the length of one side was within 1 km of another side. However, a factor to consider is the proximity of the Jovis Tholus western and eastern centres (9.20 km). Such a short base would have a biased tendency to form an isosceles triangle with another site. Ulysses Tholus, Alba Mons and Arsia Mons all created isosceles triangles with the 2 centres of Jovis Tholus. By not including these triangles, the total count of isosceles triangles for the 29 Martian sites is reduced to 17 triangles. Furthermore, another consideration that must be taken into account is that 7 of the 17 Martian isosceles triangles were measured out by using 2 of the sites in each triangle to survey for the 3rd site, and the equal sides of an 8th triangle were surveyed from its apex, so they cannot be used in a statistical test since they are biased. This drastically reduces the number of isosceles triangles which can be included in a statistical test to 9.

To determine the frequency of occurrence of isosceles triangles under random conditions, I created 10 simulations of 29 randomly located sites and determined the number of isosceles triangles for each simulation which met my criterion of ±0.5 km for equality of 2 sides. The average number of isosceles triangles from the 10 simulations was found to be 5.10 with a standard deviation of 2.64. The probability of 9 isosceles triangles occurring naturally yields a p-value of 0.0701 or about 1 in 14 which would not be considered by science to be statistically different from randomness.

Table 13.2 lists the 17 isosceles triangles found for the 29 sites. These do not include isosceles triangles created by using the 2 centres of Jovis Tholus for a base. The 1st and 3rd triangles had to be excluded from statistical testing since the vertex between the equal sides was surveyed from the vertices of the base side. The same applies to the 5th triangle (Jovis Tholus West with Olympus Mons and Ascraeus Mons) and the 11th triangle (Alba Mons with Olympus Mons and Ascraeus Mons). The sides of the 2nd triangle were both surveyed from Arsia Mons so it could not be counted. The final item in Table 13.2 is actually 3 isosceles triangles formed with 3 different sites equidistant from Apollinaris Mons. All 3 had to be excluded from statistical testing since Apollinaris Mons was surveyed from the other 3 sites.

Hence, although it is likely that there are more isosceles triangles present with the 29 major sites on Mars than would be expected under random conditions, this cannot be established statistically until an independent method of coordinate determination becomes available for triangles with a vertex surveyed from the vertices of the base side.

Table 13.2: Isosceles triangles from 29 sites on Mars

Vertex between equal sides	Vertices of base side	Length of equal sides		
		Sacred formula	Actual km	Diff. from sacred
Ascraeus Mons	AscSC1 Crater	$R/(2\pi)$	540.52	0.00
	AscSC2 Crater	$R/(2\pi)$	540.52	0.00
Arsia Mons	Biblis Tholus	$R/(3\sqrt{3})$	653.60	0.00
	Ulysses Tholus	$R/(3\sqrt{3})$	653.60	0.00
Biblis Tholus	Pavonis Mons	$R/(3\sqrt{3})$	653.59	0.00
	Arsia Mons	$R/(3\sqrt{3})$	653.60	0.00
Uranius Mons	Uranius Tholus	$eR/(12\pi)$	245.51	0.63
	Ceraunius Tholus	$eR/(12\pi)$	244.51	-0.37
Jovis Tholus West	Olympus Mons	$2\pi R/25$	853.55	0.00
	Ascraeus Mons	$2\pi R/25$	853.56	0.00
Ulysses Tholus	Olympus Mons	$2\pi R/30$	1119.21	1.91
	Ascraeus Mons	$2\pi R/30$	1120.11	2.81
Ulysses Tholus	Ceraunius Tholus	$\pi R/(4\sqrt{2})$	1884.92	-1.19
	Tharsis Tholus N	$\pi R/(4\sqrt{2})$	1885.60	-0.51
Issedon Tholus	Uranius Tholus	$\pi R/(11\varphi)$	598.75	-0.71
	Uranius Mons	$\pi R/(11\varphi)$	598.92	-0.54
Alba Mons	AscSC1 Crater	$\pi R/(3e)$	1308.90	0.54
	Jovis Tholus West	$\pi R/(3e)$	1309.19	0.84
Alba Mons	AscSC1 Crater	$\pi R/(3e)$	1308.90	0.54
	Jovis Tholus East	$\pi R/(3e)$	1308.36	0.00
Alba Mons	Olympus Mons	$R/2$	1698.09	0.00
	Ascraeus Mons	$R/2$	1698.10	0.00
Alba Mons	Ulysses Tholus	$2R'/3$	2248.62	-2.18
	Pavonis Mons	$2R'/3$	2247.93	-2.87
Pavonis Mons	Paros Crater	$Re^2/17$	1478.87	2.71
	Tharsis Tholus S	$Re^2/17$	1478.05	1.89
Ceraunius Tholus	Olympus Mons	$eR/(8\sqrt{3})$	1997.12	-1.62
	Biblis Tholus	$eR/(8\sqrt{3})$	1997.56	-1.18
Apollinaris Mons	Issedon Tholus	eR/φ	5705.30	-0.26
	Uranius Mons	eR/φ	5706.05	0.48
	Tharsis Tholus N	eR/φ	5705.34	-0.23

Testing the Distribution of Distances Between Sites

An unexpected finding was observed from the simulations of randomly located sites. Table 13.3 shows that there were statistically significant differences between the average number of distances found for some distance intervals with the 10 simulations versus the actual number of occurrences from the sites on Mars. Distances less than 1500 km and between 6000 to 7000 km occur much more frequently with the Martian sites than with the 10 simulations. On the other hand, distances between 2500 to 4000 km occur much less frequently on Mars than for the simulations. The probability of this happening under random conditions is less than 0.0000001 or 1 in 10 million for some of the distance ranges. So after 2 failures to demonstrate artificiality statistically, we now have a test which gives very strong evidence that the sites on Mars are positioned non-randomly.

Testing the Incidence of Integer (and 1/2) Longitudes and Sacred Latitudes

Having found a large number of longitudes with integer or integer and one-half values when referenced to a prime meridian such as the longitude of Elysium Mons, I wanted to check whether or not the incidence of this phenomenon was above random chance. To do this I set up the binomial expansion equation:

$$P(k) = \frac{n! \; p^k q^{(n-k)}}{k!(n-k)!}$$

where P(k) is the probability under random conditions of finding k or more integer or integer and one-half longitude values for 47 randomly located sites over a longitude range from 0 to 127 degrees. The number n is the total number of sites other than the prime meridian site (i.e., n = 47), p is the probability of having a longitude with an integer or integer and one-half value and q is the probability of not getting such a value (q = 1 - p). When I used a maximum permissible error of ±2 minutes of a degree, a total of 11 of the 47 sites qualified as having an integer or integer and 1/2 longitude value from Elysium Mons. The probability p was equal to 4 min/30 min = 0.1333 and q = 1 - p = 0.8666. The probability P(k=11) was calculated to be 0.0420 or about 1 in 24 which is considered to be just statistically significant, suggesting that this phenomenon is not likely due to randomness.

The test for the incidence of sacred latitudes with reference to the equator was conducted similarly. To keep matters simple it was done only for sacred numbers equal to π, π^2, φ, φ^2, e or e^2 expressed directly as degrees and multiplied by integers to give latitude values over a range from 9.4° S

Table 13.3: *Mean number of distances between 29 sites from 10 simulations with random conditions. The z-value is a standardized measure of the deviation of the Martian data from the simulation mean. This value is used to calculate the probability of the Martian data. The extremely low p-values found for some intervals indicate that the sites on Mars are not randomly located. SD means* **standard deviation** *and ns means* **not significant**.

Interval (km)	Mean # of distances from 10 simulations	SD	# of distances observed on Mars	z-value	p-value (2-tailed)
0 - 500	14.00	2.21	26	5.427	<0.0000001
500 - 1000	36.40	4.40	74	8.542	<0.0000001
1000 - 1500	44.40	6.88	64	2.848	0.0044
1500 - 2000	52.60	11.05	54	0.127	ns
2000 - 2500	50.40	9.34	42	-0.900	ns
2500 - 3000	47.30	4.37	15	-7.386	<0.0000001
3000 - 3500	38.30	7.48	13	-3.381	0.0007
3500 - 4000	31.70	7.02	11	-2.947	0.0032
4000 - 4500	25.90	8.25	23	-0.351	ns
4500 - 5000	20.60	7.83	11	-1.225	ns
5000 - 5500	17.60	5.27	22	0.834	ns
5500 - 6000	14.00	5.77	23	1.559	ns
6000 - 6500	8.10	3.98	18	2.485	0.0130
6500 - 7000	3.60	3.24	10	1.976	0.0481
7000 - 7500	1.10	1.10	0	-1.000	ns
Total	406		406		

to 39.5° N (a total spread of 48.9 degrees). A set of 88 possible values are present in this range. I once again used a maximum permissible error of ±2 minutes of a degree. The planetocentric latitudes for Pavonis Mons and the Pentagram Pyramid were used instead of their planetographic latitudes in order to be consistent for this test. A total of 14 sites were found to have latitude values which met the criteria. These are listed in Table 13.4 with their sacred latitude formulae. Some of the sites were on the same mountain so they can be expected to have the similar latitudes and this could produce a bias. Hence, I reduced the 2 sites associated with Olympus Mons and the 3 sites with associated with Uranius Tholus to a single site each for statistical testing leaving a total of 11 sites out of 45 sites. The probability p was (88 x 2 min x 2) / (48.9 degrees x 60 min/degree) = 0.1200 and q = 1 - p = 0.8800. The probability P(k=11) was calculated to be 0.0154 or about 1 in 65 which is considered statistically significant.

Table 13.4: *Sites with simple latitude formulae.*

Site	Latitude Formula (°)	Site	Latitude Formula (°)
Arsia Mons	5φ	Olympus Mons	$7\varphi^2$
Apollinaris Mons	e^2	Paros Crater	7π
Pavonis Mons	φ	Elysium Mons	$9e$
Ulysses Tholus SE Crater	φ^2	Uranius Tholus Caldera	$10\varphi^2$
Biblis Tholus	φ^2	Uranius Tholus Round Top	$10\varphi^2$
Tharsis Tholus South	4π	Uranius Tholus	$10\varphi^2$
Olympus Mons Central Caldera	$7\varphi^2$	Alba Mons	$4\pi^2$

So statistically, both longitude and latitudes values were found to support the artificiality hypothesis. Since latitude values should be independent of longitude values under random conditions, there is good justification for multiplying the 2 probabilities together to get an overall estimate of the probability for both events occurring randomly together. This gives us a value of 0.000647 or about 1 in 1500. The case for randomness has been dealt another solid blow.

Probability of Bearing Angles Found for Straight Line Segments in the Mountains

Another question to ask is how likely is it to obtain the results shown in Table 11.1 (see Chapter 11) for the bearing angles of the straight line segments found in the Martian mountains? For simplicity, if we can assume a resolution of 1° for bearing angle determination and pool the counter and counterclockwise bearing angles (i.e., take their absolute values), then there will be a total of 91 possible bearing angle values which could occur under random conditions. For a dataset of 140 straight line segments with random bearing angles, the binomial expansion equation calculates the probability of any angle value occurring 5 times or more as $P(k=5) = 0.0198$ or 1 in about 50. This establishes the minimum number of values required for statistical significance of non-randomness for any particular angle value in this set of line segments. The probability of any angle value occurring 17 times or more (i.e., for angles 0°, 36° or 45°) is about 1 in 2 trillion. To occur 20 times or more (i.e., for 36°) the probability is less than 1 in 20,000 trillion. For the line segments with the pentagram-related bearing angles of 18°, 36°, 54° and 72° the probability of their combined observed frequencies in this dataset occurring under random conditions is less than 1 in 40,000 trillion trillion. All of this indicates that the only conclusion which makes any sense at all is that these mountains must have been artificially constructed.

Summary and Conclusion

The presence of sacred distance formulae between sites was not strong enough evidence on its own to declare artificiality. This is due to not being able to reduce, without introducing investigator bias, the number of possible sacred formulae to only those likely to have been used to convey particular messages of sacred geometry. The test for a non-random incidence of isosceles triangles also failed to establish the presence of artificiality since many of the isosceles triangles were determined from surveying their apexes from 2 other sites, thus introducing a bias which could only be handled by dropping these triangles from the test. The number of isosceles triangles remaining for testing was too small to demonstrate a statistical difference from randomness. In contrast to the inadequacy of these 2 initial tests in demonstrating artificiality of site placement, the theory that the Martian sites selected for study were artificially located is supported by several other tests. It was demonstrated that statistically, it is highly improbable that distances between sites would have the frequency distribution which is found on Mars, that site longitudes from Elysium Mons would so often be integer or integer and 1/2 values, and that latitudes so often would be simple sacred formulae. The accuracy of the coordinates determined for several site placements will receive more support from evidence presented in the next 2 books of the Intelligent Mars series. As well, additional statistical tests will be provided which reinforce the case for artificiality.

The data from the bearing angles of straight line segments on the Martian mountains demonstrate non-randomness so powerfully that there remains little option but to declare the artificiality hypothesis for mountain construction to be established beyond a reasonable doubt. It was not simply a matter of drilling down to molten rock and letting it gush up through the drill hole to spew in all directions when it reached the surface, such as occurs in a volcanic eruption. Rather, the mountains were deliberately shaped by an unknown technology, resulting in straight line edges occurring over long distances. These lines have bearing angle values which are restricted mostly to those which are evenly divisible by 3 and relate to the pentagram, the equilateral triangle and the square or rectangle.

The unassisted mind was right after all! Much of the Martin topography has been engineered into place. Henceforth, I will simply assume the presence of artificiality, and focus on trying to further unlock the architectural puzzle on Mars.

14

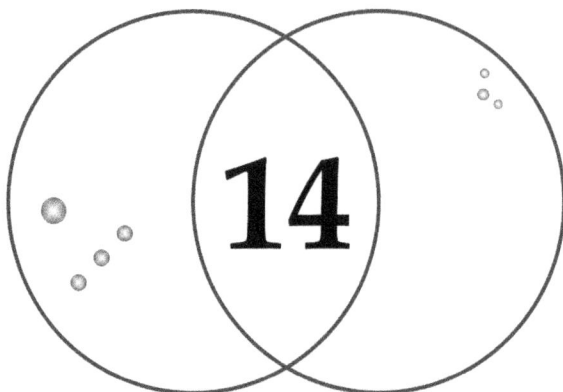

Bringing it all Together

I s there any way in which all of the findings so far can be tied together? In my quest for patterns in the layout of the Martian mountains, I came across several major themes. First of all there is R, the equatorial radius of Mars and R', the northern polar radius of Mars. All sacred distances seem to be defined in terms of R or R', thus pointing to the possibility that the architects used R and R' as standards of measure, albeit extremely large ones. The first instance in which we saw the use of R was in the bisected isosceles triangle fitting the giant mountains in Chapter 5. The locations of the triangle vertices and base midpoint were determined with a high degree of precision with survey craters which also used R as a component in their distance formulae from crater to mountain centre. It is not possible at this point to determine the exact set of distances between all the important sites on Mars which were intended to be sacred. It would seem unlikely that each site could have been placed at meaningful sacred distances from all the other sites. Nevertheless, it would appear that the architects intended to maximize this to a remarkable degree. Individual groupings of mountain and crater sites seem to be nested within an overall master blueprint which links all sites to one another. Rather than simply being scattered patterns in isolation from one another, they form one vast design which covers more than 1/3 of the planet in longitude and more than 1/4 of the planet in latitude.

Since artificiality has been established in the previous chapter, I can now reduce the number of sacred distance formulae to more manageable levels for statistical testing. If all of the 406 distances occurring between the same 29 sites used in the previous chapter are fit to the closest sacred distance formula from a subset of all possible values, 147 or 36.2% are found to be within ±1 km of a sacred distance. Since 27 of these distances were survey distances between sites, I reduced this number to 120 or 29.6% to avoid bias. I limited sacred distance formulae in the subset to those containing R together with 1 or 2 sacred irrational numbers (φ, π, e, $\sqrt{5}$, $\sqrt{3}$ or $\sqrt{2}$) or their squares (φ^2, π^2, or e^2). These could be multiplied or divided by an integer $\leq$ 20 or an integer related to the pentagram (25, 27, 36, 54, 72 and 108). Music interval ratios and their inverses multiplied by R, and the ratio 2R/5, were also permitted. I repeated the same exercise on a simulation consisting of 29 randomly located sites. Only 92 or 22.7% of these 406 distances were found to be within ±1 km of a sacred distance from the subset of values. A chi-square statistical test shows this difference to be highly significant with a p-value of 0.0009 indicating that the likelihood of this occurring by chance is only 1 in 1,100. A similar phenomena occurs when the polar radius R' is used in addition to R. In this situation, 216 or 53.2 % of the 406 distances from the 29 sites on Mars are within ±1 km of a sacred distance. Subtracting out the 27 survey distances yields 189 or 46.6%. Only 137 or 33.7% of the 406 distances from the 29 randomly located sites met this criterion. This is significant at much less than P<0.0001 or 1 in 10,000. Hence, what could not be done in the previous chapter for sacred distance formulae, has now been accomplished for a meaningful subset of sacred distances using either R or R' in their formulae. So it can now be said with some confidence that sacred distances with meaningful formulae do occur more frequently on Mars between important sites than could be expected when only randomness applies.

After R, the next most important theme seems to be the triangle, - not just any triangle, but a triangle composed of sides whose distances were all determined to be sacred, and hence itself could be called sacred. Although I came across other geometric shapes on Mars such as the pentagram in the Pentagram Pyramid and the golden rectangular shape of the perimeter of the base of Issedon Tholus, I could find no other geometric pattern in the layout of the mountains. It is true that several different polygons could be created out of sacred mountain-to-mountain distances, but none of these would appear to be meaningful. The triangle celebrates the integer 3, and forms the most rigid geometric shape. So it is a symbol of strength and has an abundance of other symbolic significance. A substantial number of the triangles were isosceles in nature. Other triangles had special properties such as 1 side being very close to 1/4 of the sum of the other 2 sides.

A third important theme is the use of the sacred irrational numbers φ, π, e, √5, √3 and √2. For instance, the bisected isosceles triangle formed from the Tharsis Montes and Olympus Mons very clearly demonstrates the use of √5 in the lengths of its base and height. The bearing angle of its base uses both e and π. Each of these 2 numbers is also utilized in distances between survey craters and the mountain centres which they survey. Isosceles triangles formed with Biblis Tholus, Ulysses Tholus and Arsia Mons utilize √3 in their equal sides. I first encountered φ in the measurement of the northern portion of the base of the triangle between Elysium Mons, Hecates Tholus and Albor Tholus (Fig. 6.4). A very clear use of √2 appears in the equal sides of the isosceles triangle formed with Ulysses Tholus, Ceraunius Tholus and the northern peak of Tharsis Tholus (Fig. 7.7). One or two of these 6 irrational numbers appear(s) to be employed in every sacred distance between sites except for those distances which can be expressed in terms of R (or R') alone, multiplied or divided by integers or by ratios of integers. They also are present in many sacred latitudes either from the equator or from Arsia Mons, and in the bearing angles of some of the rhumb lines between sites.

Each of the sacred irrational numbers φ, π, e, √5, √3 and √2 seems to represent a particular geometric figure and vice versa. Thus φ, √5 and e (see Chapter 9) would represent a pentagram and π would represent a circle such as the circular nature of a pentagram. The number √3 would represent an equilateral triangle since the tangent of each of its 60° angles is √3. The number √2 would represent a square since the diagonal length of a square is equal to √2 times the length of a side.

A fourth theme is the prominence given to certain integers. These integers are found in sacred formulae for distances and in the values of bearing angles. Like the irrational numbers, these integers seem to be used to represent geometric shapes. This is seen most strongly in the numbers referring to the angles or to the number of star points of a pentagram. Thus 5 or 25 (5 x 5) refers to the number of star points, 36 and 18 refer to the angle and 1/2 angle of a star point, 72 refers to an angle at the base of a star point and 108, 54 and 27 refer to an angle, 1/2 angle and 1/4 angle between the star points or an angle of its interior pentagon. The numbers 60 and 30 possibly refer to an angle and 1/2 angle of an equilateral triangle. The number 45, being 1/2 of 90, might refer to the 90° angles of a square, a golden rectangle and/or a double square. The golden rectangle and double square would connect 90° and 45° to φ and to √5 respectively, and therefore to the pentagram as well.

So where does all of that leave us? Although there may have been practical uses for the Martian architecture such as the pointer function of Albor Tholus and Hecates Tholus to the North Pole, much of what has been mentioned so far points to the extensive use of sacred geometry for a spiritual purpose. But what message were the architects trying to convey with the various sites and their extremely careful positioning to create the sacred geometry? The strongest symbolism appears to be associated with the planetary radius and with the pentagram. All of the sacred distance formulae utilize either the equatorial or polar radius of Mars. Four out of the six irrational numbers used can symbolize the pentagram. A large proportion of the integers used for the bearing angles of linear segments of mountain and caldera perimeters relate to angle sizes in the pentagram. The use of the planetary radius suggests that the entire planet was an integral part of the monument which they wished to create. The involvement of the planet is reinforced by the use of longitude coordinates which are in integer or integer and one-half relationship to a prime meridian, most likely Elysium Mons, and of latitude coordinates which are in terms of the 6 irrational numbers as well as integers. Perhaps the Martian civilization was reverential to the planet in much the same way we are reverential to Mother Earth. The pentagram may have been a focus of the creative aspect of the Divine since φ appears so often in Nature. The expanding or diminishing pentagram series can extend to infinity in both directions thereby mimicking the infinity aspect of Divine creation, part of which is the planet itself. All in all, we are probably looking at the work of a very spiritually motivated and spiritually evolved civilization.

But there still seems to be something missing in this explanation. Although a huge pentagram pyramid was part of the architecture, the much larger mountain monuments are not directly in the image of a pentagram. None of the sites are arranged to create a pentagram and none of the mountains have a pentagram shape. The most impressive and obvious monument appears to be the bisected isosceles triangle created by Olympus Mons and the 3 Tharsis Montes. Other than having √5 in its height and base dimensions, its connection to the pentagram appears rather weak. Although the double square is implied this is not explicitly marked out by mountain sites. To explain the most likely significance of the bisected isosceles triangle, I will now turn to a very unlikely source - the work of a highly renowned artist and inventor from the 15th &16th century, namely, Leonardo Da Vinci.

We have a splendid example of how sacred geometry can represent the human body in Leonardo Da Vinci's Vitruvian Man (Fig. 14.1). Da Vinci's drawing shows how a human body fits both a square and a circle. The circle is centred on the navel and passes well above the head. In order to touch both the square and the circle, Da Vinci has drawn 2 sets of arms and legs. In one set, the arms are angled upwards and the legs are spread apart to permit the man to touch the circumference of the circle. In the

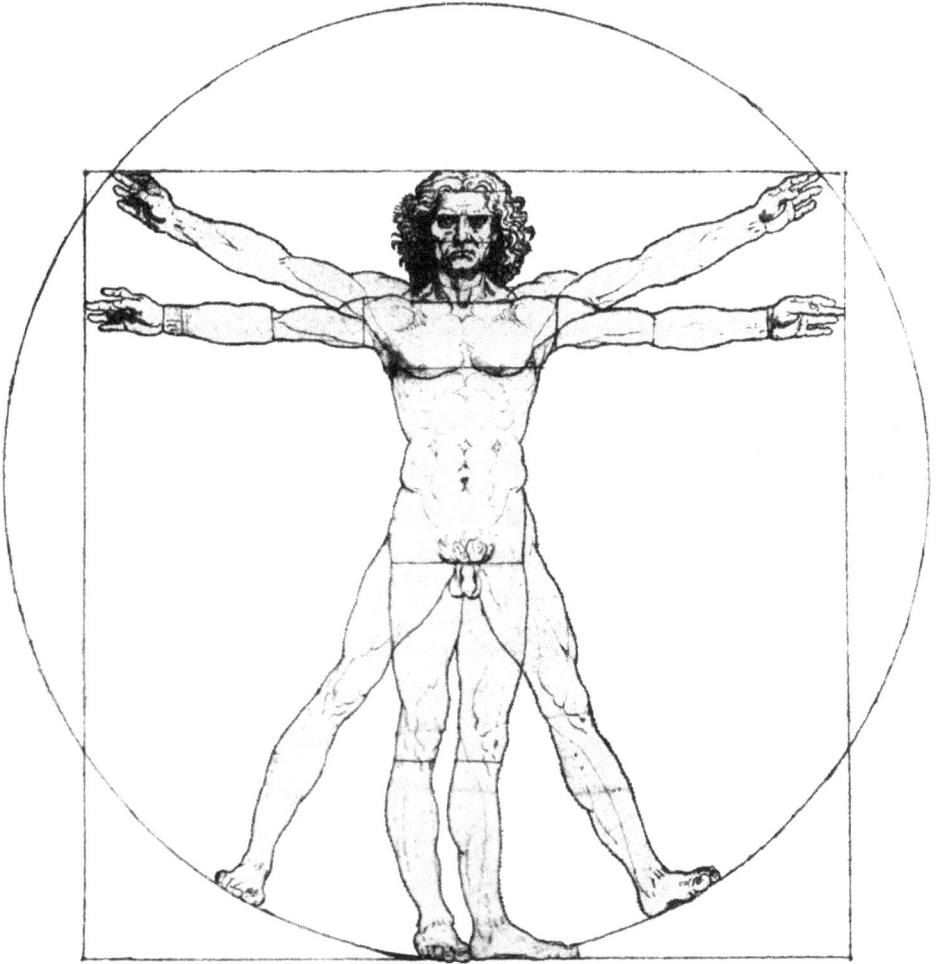

Fig. 14.1: *Leonardo Da Vinci's drawing of the Vitruvian Man created sometime between 1485 and 1490. It shows that the human body can be well fit to both a square and a circle. Photo modified from McMurrich[1].*

other set, the arms are extended horizontally to touch the sides of the square and the legs are together allowing the person to stand on the bottom line of the square. The top of the man's head touches the top line of the square. Thus the height of the square marks the height of the person when standing straight upright with legs together at the bottom of the square, and the width of the square is a measure of the arms stretched out horizontally. A square is formed because, as Da Vinci writes, "the length of the outspread arms is equal to the height of a man". So from the Vitruvian Man, we can extract the sacred geometric symbols of π from the circle and $\sqrt{2}$ from the diagonal of the square, the number 4 from the number of sides of the square, and the geometric shapes of a circle and square to represent the shape of the human body.

The Vitruvian Man becomes even more interesting from the viewpoint of sacred geometry when a vertical line is drawn from the top of the head through the centre of the square to connect with the bottom of the square between the pair of feet which are together (Fig. 14.2). This line divides the square into 2 equal portions, each of which could be further subdivided into 2 small squares to form a double square by passing a horizontal line (not shown) through the centre of Da Vinci's square. By joining each of the 2 top corners of Da Vinci's square to the point where the vertical line touches the base side, 2 lines are formed which are diagonals of the double squares. If a value of 2 arbitrary units is assigned for the side length of Da Vinci's square, the width of each double square becomes 1 and the diagonal length becomes equal to $\sqrt{5}$. A bisected isosceles triangle has been created which is exactly analogous, but upside down, to the left diagram in Fig. 4.5 in Chapter 4. Lo and behold, we have created the model for the bisected isosceles triangle formed by Olympus Mons and the 3 Tharsis Montes!

A closer look at the Vitruvian Man reveals that the 4th fingers of the arms which are extended upwards point at the 2 upward corners of the square. The use of the 4th fingers for this purpose may have been a deliberate reference to the 4 sides of the square. If we superimpose the Vitruvian Man on the 4 giant mountains, we get a picture of a human figure which fits the mountain monument very well despite the distortions arising from trying to lay a 2 dimensional figure onto a spherical surface (Fig. 14.3). The feet which are together stand very close to the survey centre for Olympus Mons and the top centre of the head approximately touches the survey centre for Pavonis Mons. The 2 upper arms reach out towards Ascraeus Mons and Asia Mons with the 4th fingers pointing at their respective survey centres.

All of this suggests a number of things. Firstly, the goodness of the fit of the geometry of the Vitruvian Man to the architecture found on Mars points to the likelihood that the mountain monument represents the geometry of the Martian body, and that the monument is based on a Martian version of the Vitruvian Man. Secondly, if the Martian architecture does represent the proportions of the Martian body, then the geometry of the Martian body must be very similar to that of the human body, creating the possibility that they might have been our ancestors. Thirdly, it is possible that Da Vinci's drawing might be code for the architecture on Mars, and that Da Vinci had access to secret documentation about Mars possibly contained in the Vatican archives or in the possession of a secret society which may have its origins in the ancient Egyptian mystery schools. Da Vinci may have even based his

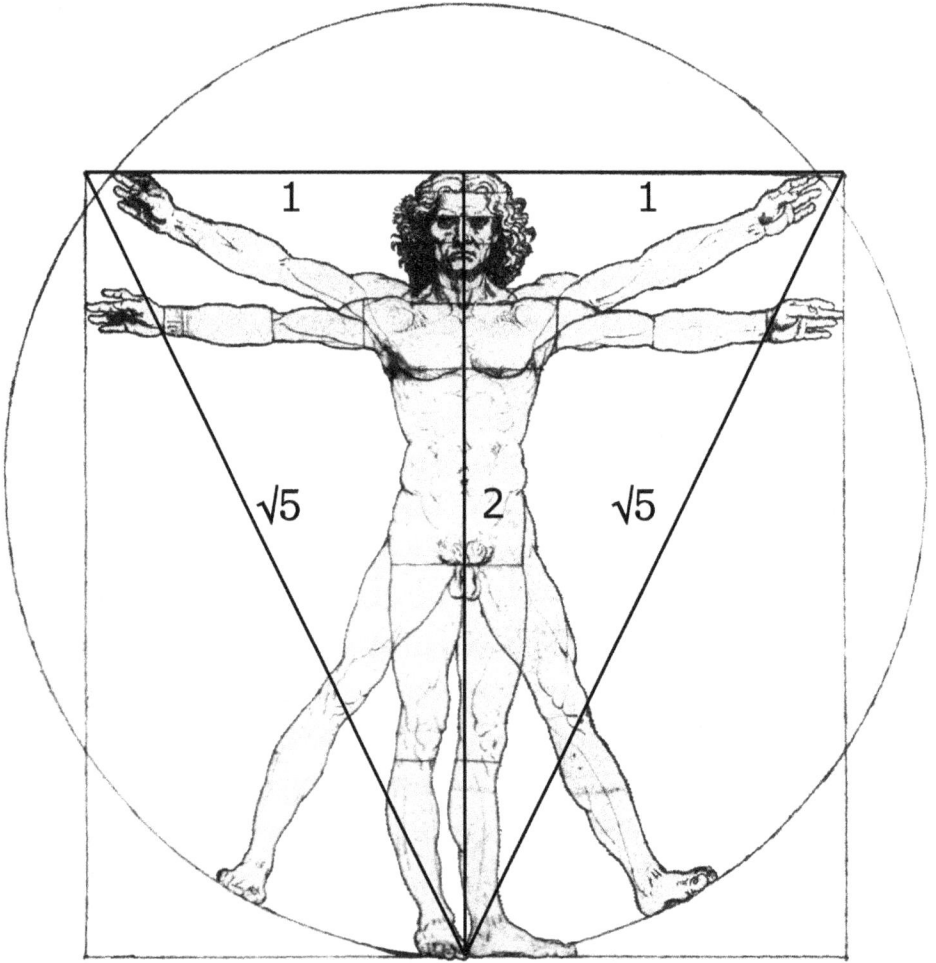

Fig. 14.2: *The hidden presence of a bisected isosceles triangle in the Vitruvian Man. It can be formed from the top line of the square as its base and the body midline as its bisector. By assigning the side length of the square a value of 2 arbitrary units, the length of each of the 2 equal sides of the isosceles triangle becomes √5 units. Vitruvian Man photo modified from McMurrich.*

drawing on a Martian version of the Vitruvian Man.

Now let's look at the irrational numbers contained in the mountain monument. From the dimensions of the base and height (R/√5 km) we obtain √5. Since the base and height of the isosceles triangle are identical, a square similar to the Vitruvian Man's square can be drawn which passes through all 4 mountain survey centres. This then represents √2 because the length of the diagonal of a square is √2 times the length of a side. The bearing angle of the base of the isosceles triangle is atan(e/π) and therefore represents e and π. The number φ can be obtained from √5 from the

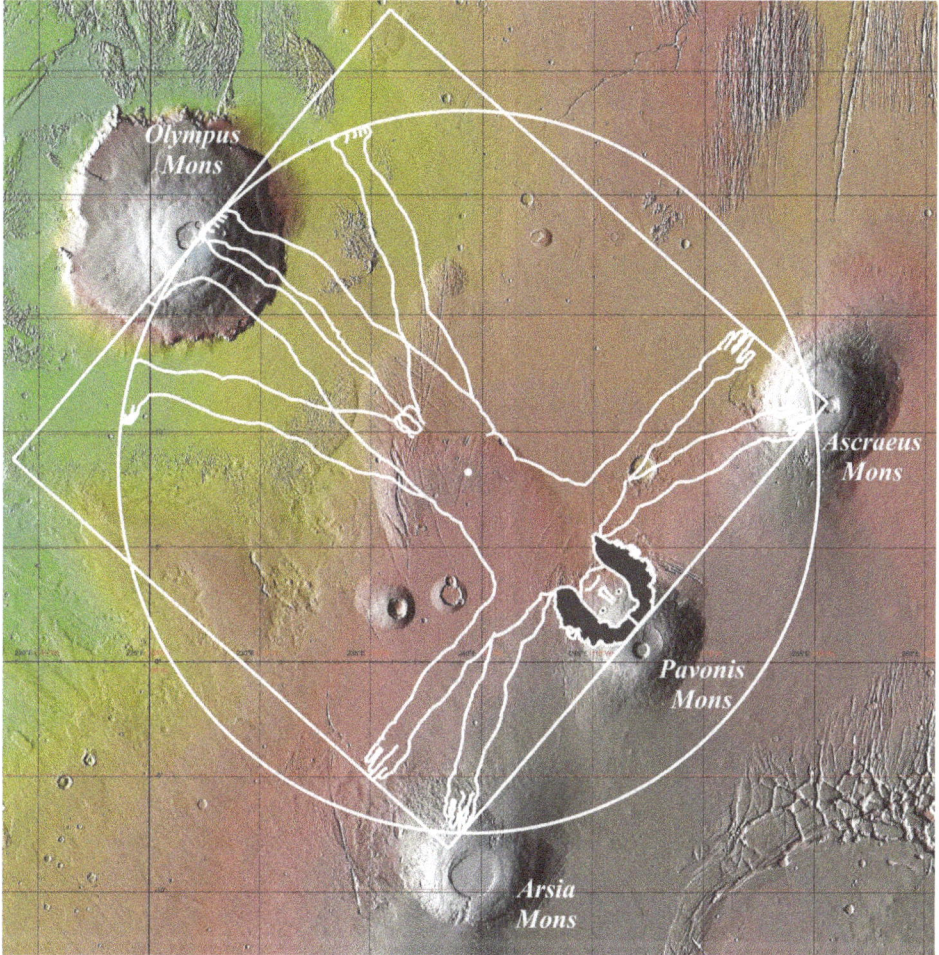

Fig. 14.3: *Outline of the Vitruvian Man overlaid on top of the Tharsis Montes and Olympus Mons. The square, the feet which are together and the top centre of the head align very closely to mountain survey centres despite some distortion due to the spherical nature of the planet. Note how the 4th fingers of the upper pair of arms point towards the Ascraeus Mons and Arsia Mons survey centres. USGS Astrogeology.*

formula $\varphi = (\sqrt{5} + 1)/2$. Thus the bisected isosceles triangle combined with the presence of the Vitruvian Man's square is even more potent than the Pentagram Pyramid in representing the irrational numbers since it only omits $\sqrt{3}$ whereas the pentagram omits both $\sqrt{3}$ and $\sqrt{2}$. However, Da Vinci wrote on his diagram of the Vitruvian Man that "if you open your legs enough that your head is lowered by one-fourteenth of your height and raise your hands enough that your extended fingers touch the top of your head, know that the centre of the extended limbs will be the navel, and the space between the legs will be an equilateral triangle". The presence of an equilateral triangle would provide a measure of $\sqrt{3}$ since the tangent of 60° is $\sqrt{3}$. If the Vitruvian Man drawing does indeed represent the Martian monument, you would expect to find some kind of marker for the circle which Da Vinci drew and the equilateral triangle which he mentions on his drawing. However, I could find no structures on Mars to identify these aspects of the Vitruvian Man. Perhaps they existed only in virtual form for the Martian monument. Nevertheless, the fact that R is used in its dimensions demonstrates that this mountain monument represents the entire planet as well. If we can assume a virtual presence of the circle and equilateral triangle, this single monument might symbolize all of the basic messages contained in the totality of the Martian mountain architecture.

Getting back to the explanation that the Martian architecture represents a high level of spirituality, we can now add that the bisected isosceles triangle of Olympus Mons and the 3 Tharsis Montes is most likely a representation of the Martian physical form. This civilization probably considered that their bodies reflected the Divine in some manner, such as is found in Christianity (e.g., St. Paul to the Corinthians [1 Corinthians 16:9]: " Don't you realize that your body is the temple of the Holy Spirit, who lives in you and was given to you by God?"). Hence, the Martian body was likely to be revered along with the planet as a very important part of Divine creation (represented by the Pentagram Pyramid). In fact, the use of R and R' in the dimensions of the monuments may have made them resonant with the whole planet and this could have a strong spiritual significance. I am thinking here that perhaps the monuments were designed to attract the energy of the Divine by being in tune with the mystical mathematical numbers embedded in the very products of creation. The architects may have intended this energy to be transferred to the planet as a whole by using the planetary radius in the monument's dimensions. This spiritual energy supposedly would protect the planet from evil and sustain life in health and abundance. And it probably worked for them for millions, if not billions, of years. Then something very wrong appears to have happened, throwing the planet out of circular orbit, stripping it of its atmosphere and surface water, and ultimately of a life-sustaining habitat. In summary, the architects were likely from a highly spiritually advanced civilization that viewed the planet as its temple and the monuments as sacred icons placed inside the temple to worship the Divine and elicit its protection.

References

1. *Leonardo Da Vinci The Anatomist (1452-1519). J. Playfair McMurrich. The Williams & Wilkins Company, Baltimore, 1930.*

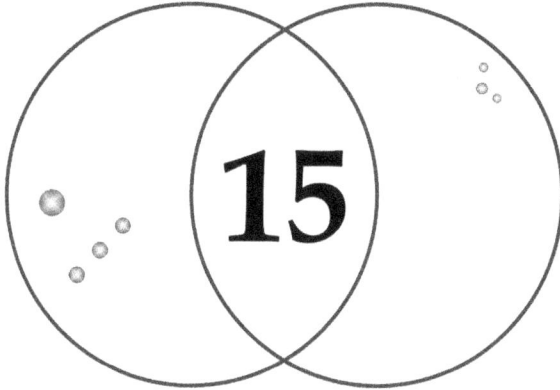

Who, When and Why?

With the discovery of this massive amount of engineering comes the obvious questions of "who did this?", "for what purpose?", and "when?". Since no one is around to tell us the answers, we must enter the realm of speculation, looking at what the data tell us and assessing everything in the light of what we know from many scientific disciplines, ancient writings and oral traditions, UFO (unidentified flying object) reports and even revelations from what might be considered to be the domain of parapsychology. Since I am not an expert in any of these, I can only offer some initial theories. No doubt many more will follow from other people with a high degree of training in specialized areas.

Theories of Why

In the last chapter, it was discussed how the mountain sites of Mars were arranged to express a spiritual message. Was this the only purpose for these structures or did they have a more practical function to perform as well? In this section, I will explore some of the possibilities.

Pressure release theory
What other purpose might the deliberate creation of gigantic mountains serve? Was it a pre-emptive plan to relieve internal planetary pressure in

an orderly way to avoid uncontrolled explosive releases as often happens on earth (e.g., the 2010 volcanic eruption in Iceland which disrupted air traffic over Europe)? Being such a technologically advanced civilization, they probably had the capability of detecting when and where eruptions would occur if left to natural forces alone. We must realize that, unlike earth, Mars has no tectonic plate movement and therefore there is no periodic volcanism at regions of crustal displacement. The tremendous pressure caused by continual radioactive heating of the interior would therefore have to be released by another mechanism which might result in far more violent releases of energy than the eruptions which we experience on earth. It has been estimated that to obtain the amount of material contained in Olympus Mons alone you would have to excavate the entire state of Texas to a depth of about 5 miles[1]! If the vast quantities of material contained in all the Martian volcanic mountains were to be stuffed back beneath its crust, the pressure required to achieve this without expanding the crust would probably be enough to violently explode the planet apart. Hence, relieving this pressure in a controlled manner makes excellent sense. And if you are going to do this, why not do it in some architecturally pleasing way instead of a hodgepodge scattering of something that will define your environment in a major way. However, in doing so, you are signalling to astute potential enemies both your presence and the extent of your technical prowess (i.e., your capacity to defend yourself militarily). Perhaps somebody did notice, and this is why the planet now appears to be so dead.

Biosphere theory

Despite the enormous volume of the volcanic material contained in the Martian mountains, it pales in comparison to the vast quantities found in the Tharsis Rise and Elysium Rise. It really makes one wonder as to why the planet had to disgorge so much of its interior. Is radioactive heating and expansion of the material in the interior really an adequate explanation to account for such high volumes? Perhaps, but I have a far more outlandish proposal. Could it be possible that the volcanic rock actually represents material deliberately excavated from a hollowing out process designed to create a habitat within the planet's interior? Let's look at some of the numbers. As mentioned in Chapter 8, I estimated the volume of Olympus Mons to be about 2.7 million cubic km. The flattened shape characteristic of shield volcanoes, with a very gentle slope to the peak of Olympus Mons (and the Tharsis Mounts) makes it very efficient to store a maximum amount of material within its radius as opposed to a more conical mountain. Another monstrous mountain is Alba Mons, the largest mountain in the solar system by area (approximately 2 million square km). Its volume has been estimated to be about 2.5 million cu km^2. By using a simple formula for the volume of a cone, I calculated the sum of the total volume of the Tharsis Montes and the Elysium group of mountains to be about 2.9 million cubic km. Thus together, the major mountains of Mars contain about 8 million cubic km of material. Then there is the volume of the Tharsis Rise estimated to be about 300 million cu km^3. The geological forces leading to construction of this vast region are still controversial. Theories as to how it arose include crustal intrusion, a mantle plume, or volcanic construction.

So altogether, there are at least 308 million cu km of volcanic material in this region of Mars. This could theoretically create a continent-sized subterranean space of about 31 million square km, or more than 3 times the size of the United States with a ceiling height of about 10 km - an altitude close to the height at which most commercial jets fly today. This is plenty of room to set up a biosphere capable of supporting many plant and animal life forms including, of course, the intelligent species of which our mountain architects and engineers were members. One possible configuration of such a biosphere is depicted in Fig. 15.1. A curved space which occupies a substantial fraction of an inner sphere of the globe could be constructed with its base at a hypothetical 50 km beneath the surface, depending on the depth of solid, unmelted rock. The crust above the space, being about 40 km in thickness and having an arched shape should be capable of supporting huge loads and stresses, e.g., meteor strikes, gravitational interactions with other planets notably Jupiter, etc. The planet's basic structure would still be held in place by the unexcavated portions of the planet and by several pillars of material spaced at regular intervals in the main body of the biosphere. A link to the outside could be established with an air-locking series of tubes and a door to the surface of the planet. The major body of the biosphere would likely be located under the Tharsis Rise, with a secondary space located under the Elysium Rise. The huge mountains themselves could represent space created due to expanding needs with the passage of time.

A wide door-like marking with 3 distinct edges at right angles to each other has already been observed near the apex on the south side of what is known as the D&M Pyramid[4]. This huge pyramid-shaped structure (about 3.8 km in length and 1.25 km in height) is located in the Cydonia region of Mars approximately 20 kilometers south of the famous face, and is named after its discoverers Vincent Di Pietro and Greg Molenaar. Cartographer Earl Torn determined that the pyramid has 5 sides and exhibits bilateral symmetry[5]. He also found that the values of $\sqrt{2}$, $\sqrt{3}$, $\sqrt{5}$, π and e appear to be encoded in the various angles and angle ratios of the structure. The eastern side of the pyramid seems to be largely collapsed,

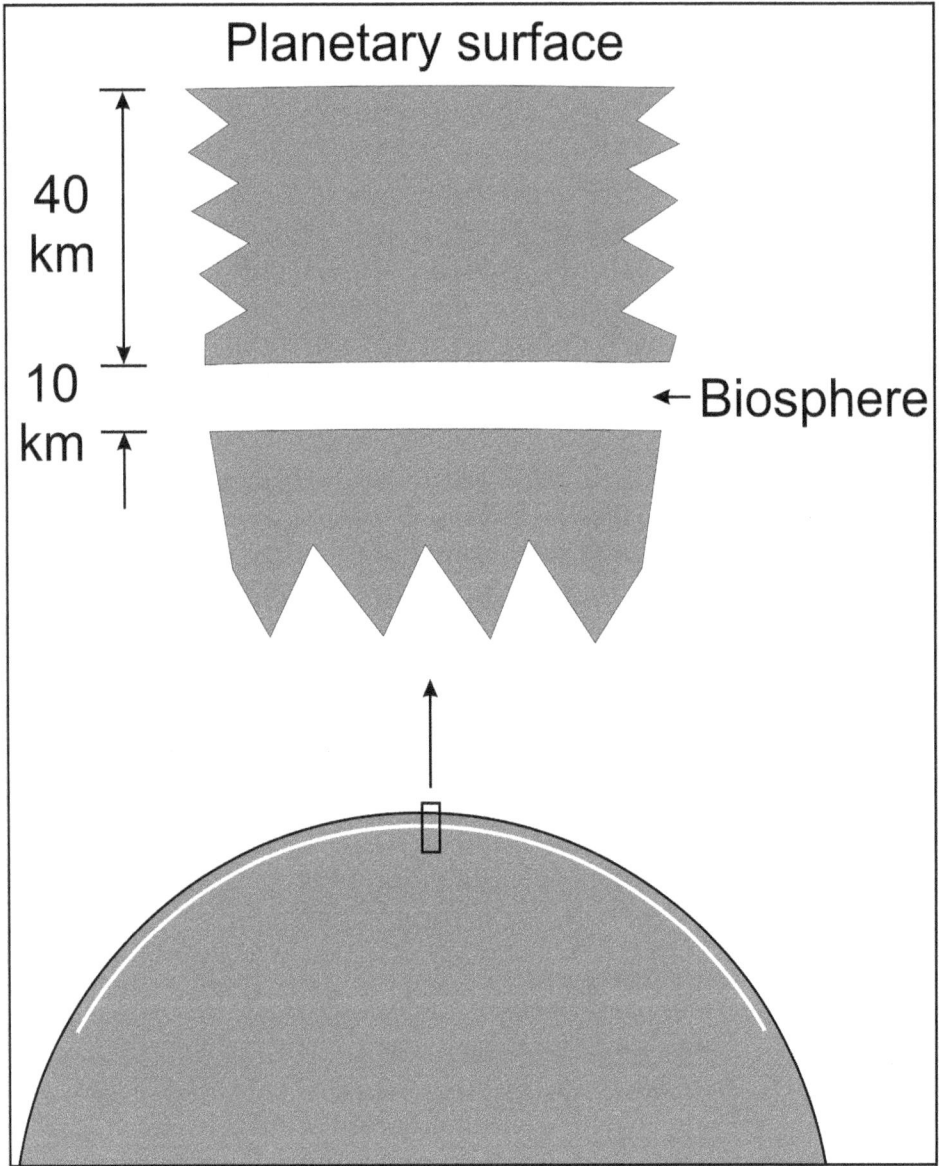

Fig. 15.1: *Lower figure is a model of an equatorial cross-section of Mars depicting a biosphere 10 km in height with its base 50 km below the planetary surface. The biosphere is shown as a curved white line which occupies 120 degrees of planetary longitude. This is about a 7000 km length at the equator. To occupy the same volume as the material contained in the Tharsis Rise plus the major Martian mountains it would have to straddle a range of approximately ±40 degrees of latitude about the equator. A magnified region is depicted in the top figure.*

suggesting a hollow interior. Hence, the D&M Pyramid complete with its postulated door is a perfect candidate for a spaceship entry port into a hypothetical subterranean biosphere. The door is approximately 90 meters square, giving plenty of clearance for a large spacecraft. Another large rectangular door dozens of feet high has been seen at the base of a large teardrop-shaped mountain in Arabia Terra[6].

An even more likely candidate for a port to the Martian underworld comes from a picture taken by the HiRISE instrument onboard the Mars Reconnaissance Orbiter of a seemingly bottomless hole on the north slope of Arsia Mons[7]. This hole is approximately 150 meters in diameter. NASA determines the hole to be at least 78 meters deep, but no bottom can be seen. Another 7 dark holes in the order of 100 meters in diameter have been photographed on a lava plain northeast of Arsia Mons[8]. They all are too dark to see anything inside. These holes are also ideally situated for entry into what probably could be the main body of the postulated subterranean biosphere, being right in the heart of the Tharsis Rise and close to the huge mountains all of which could have served as conveniently located dump sites for the excavated material. The D&M Pyramid on the other hand is well north and east of the Tharsis Rise, and west of the Elysium Rise. Maybe it is simply a spacecraft hanger rather than a port to the biosphere. The same comment is suitable for the door in the Arabia Terra mountain as it is situated far from both the Tharsis Rise and the Elysium Rise.

Excavation of the biosphere would require a heat source to liquefy solid rock, and a pressure source to drive the magma through a drilled channel and ultimately out of the top of an old or newly created caldera site. The fact that lava flows occur for incredible distances from their source vents (up to 1300 km for Alba Mons) suggests that the lava might have been heated well beyond the melting point or perhaps an external application of heating was used, such as a beam of energy, to keep the lava flowing. In the case of the Tharsis Rise and Elysium Rise, there may have been multiple calderas distributed over hundreds and even thousands of kilometers to cover the huge expanse of such landforms. The calderas may have later been sealed over on the rises since none are visible today. If the mountains were formed from magma rising up through a central channel, their original calderas may have been replaced by precisely located sculpted versions, or the original ones could have been sculpted in situ. Such possibilities would account for the straight line segments found in several mountain calderas. Volatiles such as steam and carbon dioxide generated in the heating phase of the magma within the planet could be used to generate the biosphere's atmosphere after a suitable separation and/or chemical rearrangement.

Guidance theory

It is very plausible that the mountains were intended to be viewed from outer space, several hundred kilometers above the surface of the planet, in order to be able to get any sense of the patterns they created. The survey craters would also have to be viewed from a considerable distance above in order to be useful pointers to the original mountain positions. One of the mountains (e.g., Elysium Mons) could have served as a highly visible marker of the prime meridian of the planet. Another purpose could be to point out to spacecraft the exact direction of the North Pole (e.g., Hecates Tholus together with Albor Tholus). Perhaps they acted as locators for a spaceport on land, especially useful in periods of huge dust storms which occurred from time to time. Seeing a particular pattern from outer space would give an instantaneous indication of where you are. And the size of such locators makes them easy to detect in dust storms with electromagnetic waves of appropriate frequency.

The use of survey craters to mark out the original position of the major mountains may also have been intended to locate underground structures oriented to the original mountain positions. For instance, new connecting tubes from the planetary surface to the biosphere may have had to be constructed to meet evolving requirements. Also, certain types of work to be done on the surface might impact subsurface structures, so an accurate survey would be necessary to help avoid sensitive areas, just as it is important for us to locate buried cables before beginning any excavation. Another purpose of carefully positioned mountains and survey craters might be to mark the declinations of important stars and various seasonal positions of the sun, much like the megalithic sites on earth. This would be an interesting area of investigation to pursue in future work.

Transmitter/Receiver Theory

The fact that the distances between sites are mainly some function of R and R', the planetary equatorial and polar radii, suggests the possibility that these sites could be resonant somehow with the entire planet, thereby creating a powerful transmission device. They might also be tuned to optimally receive a particular set of wavelengths. This civilization obviously came from outside the solar system, so the parent civilization had to come from another star within the galaxy or even perhaps another galaxy if they were capable of overcoming the speed of light limitations. They undoubtedly would have wanted to communicate with their source civilization, and may have found a means with these large structures. However, the communications technology, like space travel, would have to transcend the limitations of the speed of light to be useful. Our own limited technology may still consider this to be impossible, but that is probably simply due to ignorance on our part. History is full of examples where technological barriers were overcome which previously were thought to be insurmountable. A completely different transmission medium other than electromagnetic waves might have been employed.

Housing Theory
Could the mountains be vast multi-storied condominiums as suggested in Chapter 11? If they were artificially constructed, they could have been designed to create a huge living space within, complete with parks, gardens, schools, libraries, shopping areas and transportation systems. It is difficult for us to conceive of apartment units located within structures of such height and magnitude. Just servicing them with light, air, water and energy would be a daunting task. A complete biosphere would have to be created to allow plant and animal life to survive and thrive within an interior space that covers many kilometers. The size of the population that could fit in all of the mountains combined could measure in the billions even assuming their body proportions were double that of the average human.

Like a biosphere beneath the planetary crust, this space would be well protected against enemies from without, and would be heavily insulated against the harsh Martian climate. Spacecraft ports may also have been designed into these structures to permit the coming and going of space travelling vehicles through openings such as the deep dark holes which were discussed above for Arsia Mons.

Theories of When

When might any of these possibilities have occurred? If the age of the mountains could be determined, then by logical extension, this would date the efforts of our ancient architects. The smaller size Tholi and Paterae on Tharsis are estimated to have been active more than 3.5 billion years ago[9], with parts of Tharsis Tholus being dated at 3.82 billion years before present time[10]. The giant Tharsis volcanoes were believed to have originated in the Late Hesperian Period about 3.3 billion years ago[11] and the Elysium volcanoes and Olympus Mons have been dated to about 2.8 billion years ago during the Early Amazonian period. The upper few hundred meters alone of Olympus Mons have been dated to 300 - 500 million years ago[12]. Whatever the final estimates of age turn out to be, it is a staggering amount of time for us to digest since we are used to estimates of only 5 – 7 million years for the appearance of the first minimally intelligent life on Earth in the form of ancestral hominids.

Once the layout plan for the mountains was decided upon, it may have been executed simultaneously for all the mountains, in which case, the earliest age estimate for a single mountain would be the age of the remaining mountains despite what the crater counts etc. would indicate. Or the mountains might have been initiated sequentially in response to the timing requirements for whatever was needed to fulfil the purpose for their construction. Regardless, the earliest age estimate of any group of mountains arranged in a sacred geometrical pattern is a measure of the latest possible time that the civilization first arrived in the solar system. I would suggest that 3 - 4 billion years ago is a reasonable estimate.

Theories of Who

The age of the mountains being several billion years can only mean that whoever created them could not have been an evolutionary product of this solar system. There simply was not enough time for this to happen. Hence, we can only conclude that they came from outside the solar system and in order to do that, they must have had the technology for space travel. But where might they be from? Are there sufficient clues in the sites we have examined to give us an answer to this?

Unfortunately, the long passage of time since their arrival on Mars may have resulted in the disappearance of the civilization's original star system. And many of the stars which are now visible in the heavens are far too young to be considered. So I will not speculate on the star of origin other than to say that it is possible that it could be a triple star system such as the Sirius system is believed by some to be. I say this because of the extensive use of the triangle in the mountain layout scheme. However, the Sirius system and other stars like the stars in Orion's belt are too young to be considered. They are more likely to be candidates for possible recent visitations to our solar system, which may be the reason why they are celebrated in megalithic sites such as the Giza Pyramids.

The possibility exists that some useful information on the nature of the architects can be obtained from psychic techniques such as remote viewing. Remote viewing is a procedure whereby a psychic individual can sense information about a target site whose coordinates have been placed in a sealed envelope by an independent person. The coordinates can include the time dimension (past, present or future) as well as longitude and latitude. One person with a proven track record is the expert US Military remote viewer, Joe McMoneagle who claims to have a 65 to 75 percent accuracy rate with military targets[13]. During one of his sessions with coordinates of the Cydonia region on Mars provided to him by a NASA employee[14], he described being in a mile and one-half to two mile high pyramid[15]. The pyramid had very large rooms and corridors that seemed out of proportion for occupation by normal sized humans. He perceived humanoid beings there which were 12 or 13 feet tall which he described as being lanky with long arms and legs[15]. He mentions later that this was about 1 million years ago, and that these beings could have been our ancestors[15]. They eventually had to leave their home due a cataclysmic disaster which one of the beings communicated to him was a passing of the planet through a comet's tail. The description of the beings seems to fit with the enormous size of the artificial sites that are seen on Mars such as the Pentagram Pyramid and Issedon Tholus. Because McMoneagle never suspected that he would be given coordinates on another planet, his session has a high probability of producing valid information. If a remote viewer has any inkling of the target, her or his imagination is likely to contaminate the results. However, verification can only be achieved in this instance by beings who have proper historical records. The likelihood of the current existence of such beings will be considered in Chapter 16.

References

1. *Mapping Mars: Science, Imagination, and the Birth of a World. Oliver Morton. Picador, USA, New York. 2002.*

2. *Alba Patera, Mars: Topography, structure, and evolution of a unique late Hesperian–early Amazonian shield volcano. Mikhail A. Ivanov and James W. Head. Journal of Geophysical Research, Vol. 111, E09003, doi:10.1029/2005JE002469, 2006.*

3. *Ancient Geodynamics and Global-Scale Hydrology on Mars. Roger J. Phillips, Maria T. Zuber, Sean C. Solomon, Matthew P. Golombek, Bruce M. Jakosky, W. Bruce Banerdt, David E. Smith, Rebecca M. E. Williams, Brian M. Hynek, Oded Aharonson, Steven A. Hauck II. Science, 30 March 2001, Vol. 291. no. 5513, pp. 2587-2591.*

4. *A Doorway on Mars? An image of an often-studied feature seems to show a rectangle. http://archives.weirdload.com/marsdoor.html*

5. *The Monuments of Mars: A City on the Edge of Forever. 4th rev. ed. Richard C. Hoagland. North Atlantic Books. Berkeley, California. 1966.*

6. *The Daily Galaxy, Dec. 31, 2007. http://www.dailygalaxy.com/my_weblog/2007/12/nasa-image-reve.html*

7. *Astronomy Picture of the Day. Sept. 28, 2007. http://apod.nasa.gov/apod/ap070928.html*

8. *NASA Orbiter Finds Possible Cave Skylights on Mars. Mars Odyssey. http://www.nasa.gov/mission_pages/odyssey/odyssey-20070921.html*

9. *Insights into the Evolutionary History of the Martian Volcanic Constructs as Seen by HRSC. Werner, S.C.; Neukum, G.; HRSC Co-Investigator Team, 1st Mars Express Science Conference, 21 - 25 February, 2005. European Space Research and Technology Centre (ESTEC), Noordwijk, The Netherlands. p. 24.*

10. *Cycles of edifice growth and destruction at Tharsis Tholus, Mars. T. Platz, P.C. McGuire, S. Münn, B. Cailleau, A. Dumke, G. Neukum, and J.N. Procter. The Smithsonian/NASA Astrophysics Data System. http://www.researchgate.net/publication/233426405_Cycles_of_edifice_growth_and_destruction_at_Tharsis_Tholus_Mars*

11. *Visions of Mars. By Olivier de Goursac. Translated from the French by Lenora Ammon. Harry N. Abrams, Inc. New York. 2005.*

12. *Cratering Chronology and the Evolution of Mars. William K. Hartmann and Gerhard Neukum, Space Science Reviews 96: 165-194, 2001.*

13. *Interview With Joe McMoneagle by Tom Csere. http://www.firedocs.com/remoteviewing/joe/9806-pw/index.html*

14 *Mind Trek. Exploring Consciousness, Time, and Space Through Remote Viewing. Joseph McMoneagle. Hampton Roads, 1997.*

15. *The End of the Line / Sightings on the Radio with Jeff Rense Sunday, June 1st, 1997, 8:00pm - 11:00pm Pacific Time, Featured Guest Joseph W. McMoneagle. http://www.mceagle.com/remote-viewing/pub/transcripts/jr970601jm-4of5.html*

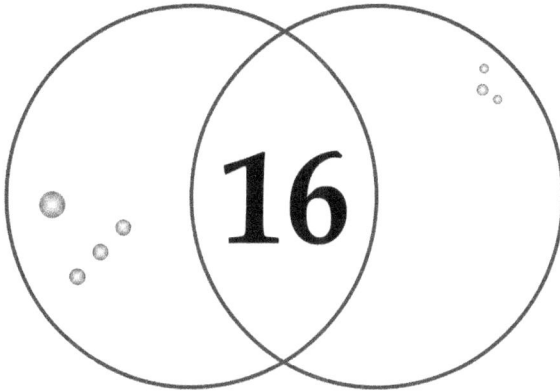

Is Mars Inhabited Today?

W hat other evidence exists that Mars was inhabited by intelligent life? And do we have any evidence to suggest that there is a continuing presence of the ancient Martian civilization, or perhaps their vanquishers, on the planet Mars? This latter question is not a totally ridiculous one even though present day Mars appears very dead according to the data and reports presented to us by NASA and the European Space Agency. However, by careful examination of satellite photos, sharp eyes and intellects have managed to pick out a large number of objects on the surface of the planet that appear to be artificial in nature. As well, we have the British and European megalithic sites whose architects used a unit of measurement based on accurate knowledge of the equatorial radius of Mars. If my hypothesis of a huge underground biosphere is correct, recent evidence of volcanism might be a sign that excavation of the interior is still occurring from time to time possibly to create more space for expansion. Then there is a large amount of anecdotal evidence which can be found in the UFO literature.

Artifacts on Mars

Faces
The most well-known artifacts on Mars are the face and D&M Pyramid both located at a region called Cydonia which lies at approximately 350.5°

E 40.8° N. They were first photographed in low resolution by the Viking missions. The face, an object about 2.6 km long by 2 km wide and 412 meters high[1], was later photographed in high resolution by the Mars Global Surveyor[2] and the European Space Agency's Mars Express[3] satellites. With the high resolution Mars Global Surveyor photograph released to the public by the Jet Propulsion Laboratory Multimission Image Processing Lab (JPL-MIPL), the face appeared to be more like a natural landform – an eroded mesa, rather than an artificial construction. However, the picture was taken from a low viewing angle which caused loss of contrast and eliminated shadows which normally bring out the facial features. Also the image was sent through low-pass and high-pass filters which had the effect of weakening image details. When these drawbacks are compensated for by Mark Kelly's expert image processing techniques, objective artificiality predictions give a combined odds against a natural formation of 1021 to one[4]. Despite these overwhelming odds, doubt persisted, especially with the release of the European Space Agency's picture of the face showing a horn on the forehead of the face[3]. A critique of this picture notes that the horn cannot be seen in any other picture taken even by the Mars Global Surveyor satellite[5]. The implication is that the picture may have been deliberately altered. Dirk Benkert, a member of the ESA High Resolution Stereo Camera team, admitted that certain scenes were "vertically exaggerated, mainly by a factor of 2 to 5 to highlight relevant surface structures"[6]. Hence, it would appear that distortions were deliberately made to make the face look more like a natural landform.

Interestingly, the coordinates of the highest point of the face (350.29864° E 40.74813° N)[1] put it at 199.9798° east of my coordinates for the north-pointing pair of mountains, Albor and Hecates Tholi. This is extremely close to an even 200 degrees, and would put it at approximately 200 + π degrees east of Elysium Mons, my candidate for the prime meridian. It's latitude is less than 2 minutes of a degree from 15e° N (40.7742 ° N).

A crowned face approximately 11 miles wide located within the Libya Montes near Syrtis Major was discovered by Greg Orme in a Mars Global Surveyor picture in 2000[7]. This huge image is rich in detail and is hard to discount as simply a natural formation. It can be divided into 3 half faces[8]. When each half face is mirrored, the half face on the left forms a crowned female face with a phoenix bird in the centre of the crown. The middle face when mirrored, produces a were-jaguar face with feline eyes, pug nose and snarling muzzle. The right half face, when mirrored, shows a moth-shaped mask with a demon mask inserted into the centre of the crown.

A colossal head was seen in a Mars Global Surveyor M.O.C. picture at about 77° E 17° N in the Syrtis Major region of Mars[9]. This head is about

19 km wide with only the upper half of the face being visible. It is very human-like with a well formed nose, closed eyes and well defined forehead, all anatomically correct in proportion. The head was not present in a previous Viking picture of the same region, suggesting deliberate tampering by NASA-JPL-MASS.

The last face which I will describe is known as the queen face. The picture of this face was taken by the Mars Global Surveyor Mars Orbital Camera and is located at coordinates 251.88° E 14.05° S. It is a female face about 780 meters wide with a long hat which by itself measures 1.3 km in length[10]. The image is somewhat faint and is not really a sculpture but rather colouration of the background rock surface. It looks very similar to Queen Nefertiti, a member of the ancient Egyptian ruling class during the time of the Pharaohs. Sceptics tried to dismiss this face as due to natural occurring shadows or to a camera artifact, but such explanations wilted when a second good picture of the same face was taken a few months later.

Then, of course, there is the 14 km long face just north of the Pentagram Pyramid which was described in Chapter 9. Unfortunately, the low resolution of the USGS map makes it difficult to make out the features of this face with any great precision, and better photographs are needed not only of the face but of the other interesting objects in this region, including the Pentagram Pyramid.

Pyramids
Pyramids have been found in great abundance on Mars. The first pyramids to be observed were found in the Elysium Quadrangle at about 162° E 15.5° N[7,11]. They were photographed by the Mariner 9 spacecraft in 1972 from an orbit approximately 1000 miles high. These pyramids are perfectly tetrahedral in shape and are truly massive, being 700 - 800 times the volume of the Great Pyramid in Egypt. The largest have a base width of 3 km and a height of 1 km. No known geological process could be found to account for these formations.

The next set of pyramids to be discovered were observed in the Cydonia region of Mars from pictures taken by the Viking spacecraft. This region contains not only the famous 5-sided D&M Pyramid described in Chapter 15, but also several pyramids in an area known as the "city" which lies to the west of the face. These include 5 major pyramids set in a pentagonal arrangement[7,12,13]. Two other large pyramidal structures lie just west of the pentagonal area. The city complex seems to offer a good view of the face. In fact, someone standing in the centre of the city complex would see the summer solstice sun rise over the face. To the east of the face lies a structure known as the "cliff" which is a straight wall 2 miles long. On the rim of a crater just to the east of the cliff is what seems to be yet another pyramid, this time a 3 sided tetrahedral one[14]. Pyramids associated with craters have also been found in other locations on Mars[7].

During his remote viewing session, Joe McMoneagle described another grouping of pyramids at coordinates 198° E 15° N which is about 1560 km west and south of Olympus Mons[15]. Pictures of these pyramids were later supplied to him by Dr. Michael Carr of the National Space Science Data Center, thus confirming the accuracy of this part of his session.

The Pentagram Pyramid will probably prove to be the most interesting pyramid on Mars. Although it is only partially visible on the USGS map, its position relative to the mountains of Mars gives it a very high degree of credibility.

Tubes

Glass-like translucent tunnels with internal opaque supporting ribs at regular intervals have been discovered on Mars[16]. There are many kilometers of them, some of them partially buried, while others have portions that are almost totally exposed. These tubes, besides running for very long distances, are extremely large in diameter, sometimes up to several hundred feet. They are thought by many observers to be transportation tunnels for vehicular traffic, possibly running in both directions.

Other Artifacts

There are numerous other artifacts on Mars which have been noted by keen observers. I will only mention a few before moving on to the next topic. One of these artifacts include a mound now known as the "Tholus" located a few miles south of the cliff in the Cydonia region[7]. The Tholus has a spiral ramp encircling it leading to the top. Then there are huge formations/discolourations which seem to be in the shape of animals at various locations on rocky surfaces. Some of these are a dolphin[17], a striped animal[9], an animal with horns[18] and a bird[19].

What looks to be animal and humanoid skulls have also been pointed out in Mars Rover pictures[20,21]. In addition, there is a huge skull-like structure or building that occurs between the face and the D&M Pyramid in Cydonia[3]. It is almost 4 km long by about 3 km wide.

Megalithic Sites on Earth

Alexander Thom examined some 600 megalithic sites in Britain and concluded that a common unit of measurement was used in these structures. He based his conclusion on the fact that the diameters of the circles and major diameters of the flattened circles were integer multiples of the same basic unit. He coined the term Megalithic yard for this unit which he determined had a value of 2.720 ± 0.003 feet (0.8291 meters). The megalithic sites are dated to have been constructed from 2000 to 1600 B.C. and thus cover at least a 400 year time span in which a precise standard unit of measure was used[22]. Using a data set of 63 measurements from sites in England and Wales, and 82 measurements from Scotland, he determined that there was no difference (i.e., greater than 0.03 inches) between the value of the Megalithic yard in England and Wales and that in Scotland. Thom therefore conjectured that this unit of measure must have been dispersed from a single place since its accuracy was consistent throughout Britain. If it had simply been copied from community to community, accumulated errors would have resulted in measurable differences. Since the Megalithic yard differs by only 0.0036 inches from $1/(2^{12} \times 10^3)$ of the equatorial radius of the planet Mars, it is reasonable to assume that the single place of origin comes from an advanced civilization on Mars itself. It also suggests that this Martian civilization used the Megalithic yard in the form of the Martian meter (see Chapter 12) as a standard unit of measurement themselves. This would indicate that a population with links to Mars was still present only 4000 years ago, and thus may have survived to the present day. Whether or not it still has a presence on Mars or has fully migrated to Earth is unknown.

Recent Volcanism

Crater counts on surfaces are an indicator of age. Volcanic flows occurring 10 million years ago or less are in evidence on the Elysium Planitia and Olympus Mons[23]. High resolution satellite pictures indicate lava flows on the large shield volcanoes as recent as 2 million years ago[24]. If the theory of an underground biosphere is correct then a recent lava flow may be evidence of further excavation of a biosphere rather than simply a natural occurrence. If so, then the lava flows would show that a Martian civilization was still active about 2 million years ago.

UFO Literature

The UFO literature is replete with references to Mars. I will only mention some examples.

Wilbert B. Smith worked with the Canadian Department of Transport in the 1950's. At the Second Storey Meeting on April 22, 1952 he said that the number of UFO sightings increased when Mars and Earth came closer together in space[25]. Project Second Storey was a group organized by the Department of National Defence in 1952 to collect and analyse UFO sightings. Wilbert Smith had top security clearance and was a key figure in Canada's UFO studies. Jacques Vallée and his spouse also found a strong relationship between UFO sighting waves and the closeness of Mars to earth[26]. However, Jacques was reluctant to proclaim a statistically valid correlation despite the seeming strength of their data.

The Curiosity Rover captured a picture of the same disc-shaped UFO in 2 separate photos[27,28]. The UFO is unlikely to be the result of dirt on the lens or missing pixels since it appears in the upper left in the first photo and in the upper centre of the second photo. A cylindrical object which is hard to construe as a spot due to a dirty lens is shown in a picture taken by the Opportunity Rover at Meridiani Planum[29]. Yet another object recorded by Spirit Rover shows a bright streak across the Martian sky which Mission controllers tried to explain away as a meteor or an old orbiting satellite rather than a spaceship as believed by many[30]. Russian Phobos II spacecraft took a picture with an infrared camera of a long ellipsoidal object interpreted to be a spacecraft very close to the Martian moon Phobos[31]. The object was 25 kilometers in length! Could this be a camera artifact?

There is also a long history of observations of strange lights or "clouds" either on the surface of Mars or emanating from Mars[32]. These observations were often reported by professional astronomers and include the sighting of "a long train of clouds" in 1862 by Sir Norman Lockyer, red lights on opposite sides of Mars in 1864 and 1865, searchlight-like beams from Mars to Earth in 1928 and 1936 by French astronomers, and brilliant transient spots of light at several different times between 1890 to 1971 by various astronomers.

Conclusion

The artifacts on Mars give the impression of being very aged, and some are in ruins due to deliberate damage or decay. They suggest that their creators are no longer active, that they somehow suffered a catastrophe or an attack that decimated them. Perhaps they vacated the planet or were taken over by a superior force. The artifacts may be so old that they were present when Mars still had an atmosphere, warm temperatures and liquid water in rivers, basins (e.g., Hellas and Argyre) and the northern ocean. If such were the case, they may be the archaeological ruins of a civilization which was eventually forced to transfer to a protected environment such as an artificial biosphere in the planet's interior. At any rate, the artifacts are very difficult to date, with speculated ages running from the near past to the very distant past.

It appears that NASA is following a policy of trying to hide or explain away the artifacts on Mars. For instance, the features of the skull-shaped structure between the face and the D&M Pyramid are absent in the Viking photos. It might be argued that the Viking photos simply lacked adequate resolution, but when the features are of the order of a kilometer in magnitude, it is hard to believe that explanation. Michael Malin, director of the Malin Space Science Systems, held on to 8 high resolution pictures of the Cydonia region taken between 1999 and 2000, before suddenly placing them on his web site, apparently in response to Senator McCain's suggestion that more rigorous oversight of NASA should be employed[33]. Then there were what appear to be deliberate distortions to recent high resolution pictures of the Cydonia face as mentioned above.

Since the artifacts are so difficult to date, some of the data on UFO's, observations by professional astronomers of lights on Mars, and the use of the Megalithic yard on planet earth give the best evidence for the possibility of a continued presence of some kind of civilization on Mars. Recent Martian volcanism might also be a product of current intelligent life if the biosphere theory is correct.

References

1. *What the Mars Global Surveyor MOLA Reveals about the Mars Face (and what it reveals about JPL). Lan Fleming.* http://www.vgl.org/webfiles /mars/face/mola/facemola.html

2. *Mars Orbiter Camera Views the "Face on Mars". Malin Space Science Systems.* http://mars.jpl.nasa.gov/mro/mgs/msss/camera/images/4_6_face_release /index.html

3. *Cydonia - the Face on Mars. ESA Mars Express. September 21, 2006.* http://www.esa.int/Our_Activities/Space_Science/Mars_Express /Cydonia_-_the_face_on_Mars

4. *Proof That the Face On Mars is Artificial. Tom Van Flandern, Meta Research. [Reprinted from the Meta Research Bulletin 2000/06/15].* http://metaresearch.org/solar system/cydonia/proof_files/proof.asp

5. *Posthuman Blues. Mac Tonnies Blog site. September 27, 2006.* http://posthumanblues.blogspot.ca/2006/09/much-ado-over-new-esa-mars- face-images.html

6. *The Two Faces of ESA.* http://spsr.utsi.edu/news/ESA3.pdf

7. *Mars Face / Inca City / Elysium Pyramids / Utopia Faces. New results 2006/2007. New Findings of the Anomalous Structures on Mars at First Found by Mariner 9 and Viking 1. New Findings of the Anomalous Structures on Mars. Walter Hain. http://www.saeti.org/index6.htm*

8. *The Crowned Face(s). George J. Haas. http://herotwins.hypermart.net/Crowned/CrownedFace.htm*

9. *Colossal Head and Animal Monuments. Report #003 by Joseph P. Skipper, June 27, 2003, Mars Anomaly Research. http://www.marsanomalyresearch.com /evidence-reports/2000/003/colossal-monuments.htm*

10. *http://marsartifacts.tripod.com/queenn.html*

11. *The Sagan Pyramids - also known as Mariner 9 4205-78 & 4296-23. http://mars-news.de/mr9/4205-78.html*

12. *An Overview of the Symbolic and Artistic Implications Surrounding the Face on Mars. Kynthia. http://www.enterprisemission.com/cydonia.html*

13. *Face it; it's a Face - (The Sequel) More on Mars Monuments. Mike Bara. http://bibliotecapleyades.net/marte/esp_marte_37.htm*

14. *Mirror of Cydonia: A Mars/Earth Connection. David Percy. http://www.aulis.com/mars.htm*

15. *Mind Trek. Exploring Consciousness, Time, and Space Through Remote Viewing. Joseph McMoneagle. Hampton Roads, 1997.*

16. *Opening a Martian "Can of Worms …?". Ron Nicks. http://www.enterprisemission.com/can.htm*

17. *The Cydonia Codex: Reflections from Mars. George J. Haas and William R. Saunders. Frog Ltd. c/o North Atlantic Books, p. 155, 2005.*

18. *Artificial Structures on Mars. Tom Van Flandern. Meta Research, Washington, D.C. April 5, 1998 - May 8, 2001. http://metaresearch.org/solar%20system/cydonia/asom/artifact_html/*

19. *Bird Catching a fish. http://www.flickr.com/photos/10805619@N02/1607170805/*

20. *Mars Humanoid Skull? Report #102 by Joseph P. Skipper, May 9, 2006, Mars Anomaly Research. http://www.marsanomalyresearch.com/evidence-reports/2006/102/mars-humanoid-skull.htm*

21. *Scary Martian Skull in Mars Curiosity Rover Photo! http://www.youtube.com/watch?v=wtH4e8mneQE*

22. *Megalithic Sites in Britain. Alexander Thom. Oxford University Press, 1979.*

23. *Cratering Chronology and the Evolution of Mars. William K. Hartmann and Gerhard Neukum, 2001. http://www2.ess.ucla.edu/~nimmo/ess250/hartmann.pdf*

24. *Recent and episodic volcanic and glacial activity on Mars revealed by the High Resolution Stereo Camera. G. Neukum, R. Jaumann, H. Hoffmann, E. Hauber, J. W. Head, A. T. Basilevsky, B. A. Ivanov, S. C. Werner, S. van Gasselt, J. B. Murray, T. McCord & The HRSC Co-Investigator Team. Nature 432, 971-979 (23 December 2004).*

25. *The UFO Files. The Canadian Connection Exposed. Palmiro Campagna. Stoddart Publishing Co. Limited, Toronto, Canada, p. 54, 1997.*

26. *Challenge to Science: The UFO Enigma. Vallée, Jacques, and Janine. Henry Regnery Company, Chicago, Illinois. 1966.*

27. *UFO Moves Across Two NASA Rover Photos On Mars, Check NASA Link to Confirm Sighting. Aug. 2013. http://www.ufosightingsdaily.com/2013/08/ufo-moves-across-two-nasa-rover-photos.html*

28. *Nasa Rover Photographs Disc-Shaped UFO Above Mars. Drishya Nair. August 25, 2013. http://www.ibtimes.co.uk/ufo-sighting-mars-nasa-image-disc-shaped-501271*

29. *Opportunity: Navigation Camera: Sol 323, Mars Exploration Rover Mission. http://marsrovers.jpl.nasa.gov/gallery/all/1/n/323 /1N156861737EFF4000P0605L0M1.HTML*

30. *UFO streaks through Martian sky. Dr David Whitehouse. BBC News, March 18, 2004. http://news.bbc.co.uk/2/hi/science/nature/3520636.stm*

31. *A Soviet "Close Encounter" by Don Ecker. UFO magazine Vol. 7 Issue 01, 1992. http://www.angelfire.com/zine/UFORCE/page72.html*

32. *Architects of the Underworld. Unriddling Atlantis, Anomalies of Mars, and the Mystery of the Sphinx. Bruce Rux, 1996, Frog, Ltd., c/o North Atlantic Books, Berkeley, California.*

33. *Who's Really Running NASA? Richard C. Hoagland, 2000. http://www.enterprisemission.com/whosnasa.html*

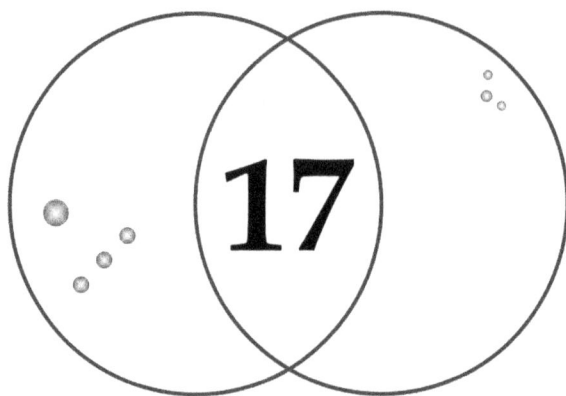

What are the Implications?

I f we are forced to conclude that a highly advanced civilization existed on Mars approximately 3 - 4 billion years ago, then we must be open to the likelihood that the planet earth was also colonized this long ago. It would be hard to believe that a civilization capable of space travel and advanced technology would ignore a planet of such beauty and high degree of compatibility for life in the form of plants and animals. I have already alluded to the link of the Martian meter to the Megalithic yard used in European megalithic sites. Cremo and Thompson have chronicled much fossil and artifactual evidence of intelligent inhabitation of the planet Earth extending hundreds of millions of years and even billions of years ago[1]. A fossilized partial shoe print complete with details of stitches was found in Triassic rock from Nevada. The rock is estimated to be 213 - 248 million years old. A gold chain was discovered in a lump of coal taken from a mine in Illinois. The coal was found by the Illinois State Geological Survey to be 260 - 320 million years old. In Dorchester Massachusetts, a metallic vase was blasted out of Precambrian rock which is over 600 million years old. Drawings inlaid with silver adorn the side of the vase. As a final example from Cremo and Thompson, a metallic sphere with 3 parallel grooves around its equator was found in a South African mineral deposit which was formed about 2.8 billion years ago, close to the same time frame proposed for the age of the major Martian mountains.

With the evidence that we now have of an advanced civilization existing at least 3 billion or so years ago on Mars, we must certainly question our current notion of the evolution of life and intelligent beings on our own planet. Importation of life from outside the solar system becomes a very real possibility. For instance, science has never been able to adequately explain the Cambrian explosion with the sudden appearance of a wide diversification of animal life with no antecedents about 530 million years ago. Now we can easily visualize the creation of a complete flora and fauna by a technologically advanced race. We ourselves might be the descendants of an extraterrestrial race. The fit of the Vitruvian Man to Martian architecture presents the possibility that our parent extraterrestrial race could have originated on Mars itself. Or perhaps we might have been the product of genetic engineering as has been suggested by Zecharia Sitchin in his interpretation of ancient Sumerian writings[2]. In this scenario, astronauts from an undetected planet which has a highly eccentric orbit bringing it into the inner solar system only every 3600 years, crossed their own DNA with that of Homo erectus to fashion our human species to work for them in mining for gold. Regardless of how many of his ideas will stand the test of future data, Sitchin's theories point out the very real possibility that modern Homo sapiens sapiens could have been the result of genetic manipulation by an advanced species of beings from either his postulated planet or elsewhere such as Mars itself.

But what about the members of this super race from Mars? Where did they come from and how did they come into existence? Did it take them 4 billion years to evolve from a microbe as is postulated by evolutionary theory for intelligent life on this planet? The first appearance of microbial cells on earth is thought to have occurred more than 3.8 billion years ago based on the ratio of C^{13} to C^{12} in ancient rocks found at Akilia, West Greenland[3]. More direct evidence for microbes on earth are the 3.3 - 3.5 billion year old fossils of cyanobacteria colonies found in western Australia[4]. These early cyanobacteria obtained their energy from photosynthesis and were thought to play an important role in bringing oxygen into the earth's atmosphere. So if an advanced civilization existed on Mars 3 - 4 billion years ago, it is credible to assume that they took at least 4 billion years to evolve to this stage from a single celled life form. This would put their own microbial origins to about 7 - 8 billion years ago. The age of the universe at this time would be only 5.7 - 6.7 billion years assuming the universe's present age to be 13.7 billion years as estimated from the data obtained with the Wilkinson Microwave Anisotropy Probe (WMAP) satellite[5]. But how long does it take for a single cell microbe to evolve? Science doesn't have a good answer for that question yet. Some scientists feel that the time available on earth was too short since the earth is estimated to be 4.5 billion years old, leaving at most only about 700 million years to develop the first microbe by 3.8 billion years ago. They therefore suggest that life may have arrived on earth from outer space as first put forward by Anaxagoras in the 5th century BC. Modern versions of this idea, known as the panspermia theory, propose that microbes could have arrived on earth by hitchhiking a ride on comets or meteors.

But the question remains. How long does it take to go from atoms and molecules to a single celled organism? One possible way to bracket this would be to determine when the universe had the starting materials and a suitable environment to begin the process. The nuclei of hydrogen, deuterium, helium, and lithium were created within 3 minutes of the big bang but it took 380,000 more years before electrons were cool enough to bind to the nuclei to form atoms. The average temperature was down to 3000 degrees Kelvin by that time, much too hot to launch any life creation. Besides, the heavier elements had yet to be created. This required star formation. Carbon, nitrogen, oxygen, phosphorus, iron, and other elements needed for life probably developed very early on with the formation of massive stars that had short lives which ended in supernova explosions, sending the first heavy elements out into space where they were incorporated into other stars and primitive gaseous planets. A planet which formed around a sun-like star 13 billion years ago has been recently discovered[6]. This planet is probably a gas giant with 2.5 times the mass of Jupiter. However, it is unlikely to be able to support life since it is expected to have only small amounts of elements such as carbon and oxygen due to its formation so early after the big bang, i.e., there would have been very few heavy elements cast out by supernova explosions at this time. By the end of the first billion years or so of the universe, there likely would have been a sufficient amount of the basic elements required for life, but it probably took several more billion years before conditions (e.g. temperature, availability of liquid water, low radiation levels and environmental stability) could have been conducive to support the actual evolution of life. When all of this would have been in place is still unknown. If we assume that it would take at least 4 billion years to develop the first unicellular ancestor of the Martian civilization, then the process of evolving this from elements and molecules would have had to begin only 1.7 - 2.7 billion years from the big bang.

So, in summary, the evolutionary timeline of life for the Martian civilization would be something like this:

Time = 0.0 years-------------Big bang
Time = 2.2 billion years-----Molecules needed for life are now present
Time = 6.2 billion years-----First unicellular species
Time =10.2 billion years-----First intelligent Martian ancestor
Time =10.3 billion years-----Martian ancestor civilization travels to Mars
from outside the solar system.

Much of the plausibility of this timeline hinges on the unknown conditions 0 – 6.2 billion years post big bang. Was there a protected sanctuary for molecules such as RNA and DNA to be created without them being blown apart by one of the numerous supernova eruptions that occurred during the early universe? It also makes the huge assumption that life can arise from chemistry and fortunate circumstances alone. Perhaps. But maybe a conscious intelligence is also needed to direct the proceedings, much like a conscious observer is needed in quantum physics to commit the uncollapsed wave function of a quantum particle to a discrete state.

Was Life Imported Into the Universe?

Whatever the case, if highly intelligent life were present 3 - 4 billion years ago, we must now be open to the possibility that it was present even much earlier. So early in fact that there was simply not enough time to have it evolve in this universe. The mind-boggling possibility which is now open for consideration is that life did not evolve originally in this universe but was actually imported from another universe! How this could be accomplished is for physicists to speculate, but it is less daunting for us to conceive than previously thought. The concept of wormholes for inter-universe travel is now given serious consideration by reputable scientists[5]. The fundamental question of when and how life was first created might then be pushed back not only billions of years, but perhaps even trillions of years!

The alternative proposal of a supreme Creator, a primary Intelligence, the ultimate Being, an underlying omnipresent Life Energy, as opposed to random evolution from solely physical starting points, is still on the table. Only our conceptual framework has changed.

Other Intelligent Life Forms

Since we now have evidence of at least 2 species of intelligent life, human terrestrials and ancient Martians, we cannot avoid thinking that there must be even more – much more than perhaps we could even begin to comprehend. The Martians had to have come from outside the solar system. That probably means that they have a parent civilization elsewhere in this galaxy or perhaps even another galaxy. Now we are up to 3 civilizations. It is not difficult to imagine that there could be a whole lot more than this in our own galaxy alone. In fact, very likely, the universe is teeming with life at all levels in a myriad of solar systems nested within a myriad of galaxies. We are solitary not in reality, but only in lack of awareness!

Martian Geography

With the revelation of the artificiality of major mountains and certain craters on Mars, we are suddenly placed in a position of having to look at the rest of the planet to determine what other geological features might be artificially engineered. There are many more sites that need to be assessed and I have done just that with the next 2 books in this series. *Intelligent Mars II* presents remarkable findings for a large number of craters on Mars and reveals much more about the organization of the Martian terrain in the light of new discoveries concerning Martian coordinate systems. *Intelligent Mars III* is devoted to exploring a seemingly insignificant crater whose influence extends far beyond its boundaries. As well it examines several large scale artifacts including a possible replica of the very architect who designed the layout of the artificial Martian topography.

The discovery of several accurate and ancient pointers to the North Pole leads me to conclude that the position of the poles on Mars has remained stable for at least about 3 billion years, the approximate age of the Hecates and Albor Tholi north-pointing pair[7]. This conflicts with the theory of polar wandering on Mars presented by Schultz[8] in 1985 which postulates that the Martian poles have moved several times over the course of its history, with the final shift occurring during the last billion years.

Is it Possible to Find Out Our Real History?

An early advanced intelligence on Mars together with the importation of the Martian meter to the planet earth strongly suggests that a record of our own origins as a species plus all of our history to date might exist on Mars today, or be retained by a secret group of their descendants or conquerors here on planet Earth. Thus the possibility arises that we might be able to find out about our own origins from a Martian-related civilization whose remnant still exists provided that they would be willing to share that information with us. Are we the result of genetic engineering, evolution, or simply devolved descendants of an extraterrestrial civilization who have forgotten their heritage and technology due to some catastrophe? Who has been managing planet Earth in the last few millennia – humans only or extraterrestrials covertly? When and how did life originate in the universe? Where did the original inhabitants of this solar system come from and how did they relate to us?

Certainly, we have to abandon the concept that mathematics and geometry were originally developed on this planet first in rudimentary form by prehistoric peoples and then more fully by civilizations such as the Egyptians, Sumerians, Babylonians and Greeks. Mars beats them all - not by mere centuries or millennia but by billions of years.

This brings us to the consideration of the history of symbolism found in our various religious groups and secret societies. The discovery of symbolism on Mars that is promoted here on Earth by Freemasonry, the Illuminati, and Satanism is eerie to say the least. The pentagram and the triangle are among the most powerful symbols used by these organizations seemingly for dark purposes. Also the whole concept of sacred geometry is a big part of their ritualism. Could it be that there is some sort of connection of the Illuminati and Freemasons to Mars? These organizations may have had privileged information about Mars from very ancient times as is possibly contained in Da Vinci's drawing of the Vitruvian Man. Leonardo Da Vinci is thought by some to have been a Freemason or member of a secret society antecedent to Freemasonry.

Massive Deception?

An inescapable implication of the findings of this book is that we may have all been massively deceived by those in charge of our "public" space programs such as NASA and the ESA. These programs are run from public funding, but rather than serving the public, they seem to be catering to a select group of individuals who are running the show covertly. How did they get this power in the first place? Why are they not accountable to the people who pay for their projects?

A further major implication of finding evidence of advanced intelligent life on another planet 3 - 4 billion years ago is that the planet Earth must be well known by other intelligent life forms in this galaxy, and probably in many regions of the universe. It is unlikely that whoever populated Mars near the beginning of the formation of our solar system were the only ones to know about our own planet. So the obvious conclusion is that off-planet intelligent life forms must have been keeping their presence hidden from us for one reason or another. Or they may simply have not shared this knowledge with the entire human race - just with a privileged few. This of course unleashes a whole host of further pressing questions. Why might they have kept it secret? And if they have revealed it to some, just who are the privileged few and why have they been selected rather than others? Does the deception extend into other areas such as religion, science, education and government? Is Earth actually being run by off-planet alien rulers who operate through on-planet proxies who are either hidden aliens, alien-human hybrids (e.g., " when the sons of God came in unto the daughters of men, and they bare children to them, the same became mighty men which were of old, men of renown." Genesis Chapter 6 Verse 4) , or simply complicit human beings serving an agenda that does not involve the rest of us? Zecharia Sitchin tells us that the Sumerian writings record the passing of rulership from alien beings to humans[9]. And despite having to face wilting ridicule, David Icke, British writer devoted to researching out who really controls the planet, maintains that Earth has been ruled in secret for thousands of years by a race of reptilians and reptilian-human hybrids[10]. Perhaps the secrecy of Freemasonry and its elite Illuminati leadership is ultimately about some form of alien control of our planet. It is no longer wise to simply ignore such theories as being too incredible. It is extremely important to keep an open mind as we struggle towards trying to uncover the truth, no matter how uncomfortable or shocking we may find it to be.

Our Future

What will be expected of us as we come into awareness of and open contact with other intelligent species in the galaxy and universe? If we can't measure up, we may have to be quarantined. As a matter of fact, it is possible that we are being quarantined right now. Perhaps our destiny will depend more on how we evolve ethically and spiritually than technologically!

References

1. *The Hidden History of The Human Race: Major Scientific Coverup Exposed. First Edition. Michael A. Cremo and Richard L. Thompson, Bhaktivedanta Book Publishing, Inc., Los Angeles, California. 1999.*

2. *The 12th Planet (Earth Chronicles, No. 1). Zecharia Sitchin. New York: Harper, 1976.*

3. *Scientists Strengthen Case For Life On Earth More Than 3.8 Billion Years Ago. University Of California, Los Angeles (2006, July 21). ScienceDaily. Science News. http://www.sciencedaily.com /releases/2006/07/060721090947.htm*

4. *Early Archean (3.3-billion to 3.5-billion-year-old) microfossils from Warrawoona Group, Australia. JW Schopf and BM Packer. Science, Vol. 237, no. 4810, pp. 70–73, 1987.*

5. *Parallel Worlds: A Journey Through Creation, Higher Dimensions, and the Future of the Cosmos. Michio Kaku. First Edition. Doubleday, New York, 2005.*

6. *A Young White Dwarf Companion to Pulsar B1620-26: Evidence for Early Planet Formation. Steinn Sigurdsson, Harvey B. Richer, Brad M. Hansen, Ingrid H. Stairs, Stephen E. Thorsett. Science 11 July 2003: Vol. 301. no. 5630, pp. 193–196.*

7. *Cratering Chronology and the Evolution of Mars. W.K. Hartmann and G. Neukum. Space Science Reviews 96, 165-194, 2001. Reproduced in: Visions of Mars. By Olivier de Goursac. Translated from the French by Lenora Ammon. Harry N. Abrams, Inc. New York. 2005. P. 156.*

8. *Polar Wandering on Mars. Peter H. Schultz. Scientific American 253, No. 6, 94-102, 1985.*

9. *The Wars of Gods and Men (3rd Book of Earth Chronicles). Zecharia Sitchin. Avon Books. 1985.*

10. *The Biggest Secret. The Book That Will Change the World. David Icke. Bridge of Love Publications USA. 1999.*

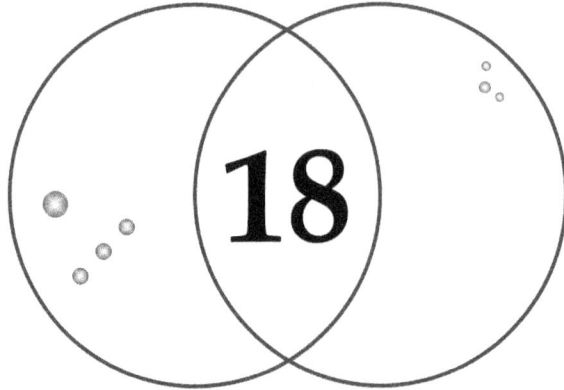

The Missing Key to the Vitruvian Martian and Concluding Remarks

Regardless of all the success I had in uncovering much of the sacred geometry on Mars, I was still left with a couple of major loose ends. Firstly, despite the Pentagram Pyramid being adjacent to the bisected isosceles triangle monument, I could not find any way in which these 2 highly important constructs could be integrated to explain their close proximity to each other. Secondly, although the drawing of the Vitruvian Man pointed to the good possibility that the bisected isosceles triangle monument represented the fit of the Martian body to a square there was no marker for a fit of the Martian body to a circle. Did the circle not work for the Martian body or did the architects simply not include it in their monument? Or did it have only a virtual presence or a presence indicated by markers too small to be visible at the resolution of the map? There was also mention of an equilateral triangle by Da Vinci in the spread-apart pair of legs but no evidence of this on the Martian terrain. I was all prepared to go to press with the wealth of material that I already did have when, at the last moment, it struck me that the Pentagram Pyramid might be acting as a key to locating the other components of the Vitruvian Man in the bisected isosceles triangle monument. But why use a pentagram? Why not a circle or a triangle? If my theory was correct, one of the components must also be a pentagram. Attempts to fit the human body to a pentagram are not new. Henri Cornelius Agrippa produced a drawing of such a fit in his 16th century work *Three Books of Occult Philosophy*.

Before I could attempt to determine whether or not the Pentagram Pyramid served as a key to locating the hidden geometrical components of the Martian equivalent of the Vitruvian Man, I needed to construct a model of how the body could be simultaneously fit to a square, a circle, a triangle and a pentagram. Leonardo is believed to have based his drawing on the writings of the Roman architect Vitruvius who stated that the distance between the tips of the fingers of a man's outstretched arms is the same as his height and that therefore a man can be inscribed in a square[1]. Vitruvius also wrote that the tips of the fingers and toes of the limbs, when spread apart, would touch the circumference of a circle whose centre was the navel. Leonardo himself stated in the writing on his Vitruvian Man illustration, that in order to touch the navel-centred circle, the middle fingers of the upward angled arms had to be at the level of the top of the head, and the legs spread out to reduce the stature by one-fourteenth. When the legs were thus spread apart, Da Vinci wrote that they would create an equilateral triangle with the navel at its apex. As for the pentagram, I made the assumption that 1 star point would touch the top of the head and a 2nd star point would coincide with the equilateral triangle vertex at the sole of the right foot of the spread-apart pair of legs.

With these starting assumptions, I set the distance between the navel and the top centre of the head to an arbitrary integer value of 1 and proceeded to calculate the dimensions and layout of the pentagram, circle, equilateral triangle and square using the principles of geometry and trigonometric functions (see Fig. 18.1). Remarkably, all 4 geometrical shapes could be integrated into a single figure. The 2 lowest star points of the pentagram coincided with the lower vertices of the equilateral triangle. The triangle sides and circle radius of my model are exactly equal to the length of a star point side plus the length of a star point base. Thus the triangle side and circle radius are in φ relationship (1.6180 times larger) to the side of a star point, and the length of any of the 5 pentagram lines is in φ relationship to either the circle radius or a triangle side (refer to Figs. 9.2 & 9.8 and text in Chapter 9). In this arrangement, all 3 geometric shapes can be considered to be an expression of φ, π, e, √5, and √3 since they are combined as a single geometrical construct in which they are measurable in units of one another. Only the square's dimensions cannot be expressed as an integer or irrational number multiple or fraction of the pentagram dimensions.

Once my theoretical Vitruvian Man was completed, I proportioned everything so that the circle overlaid that in the Da Vinci drawing (Fig. 18.2). The drawing used was taken from a book published in 1930[1]. Like all photographic reproductions, it had to be first corrected for lens distortions.

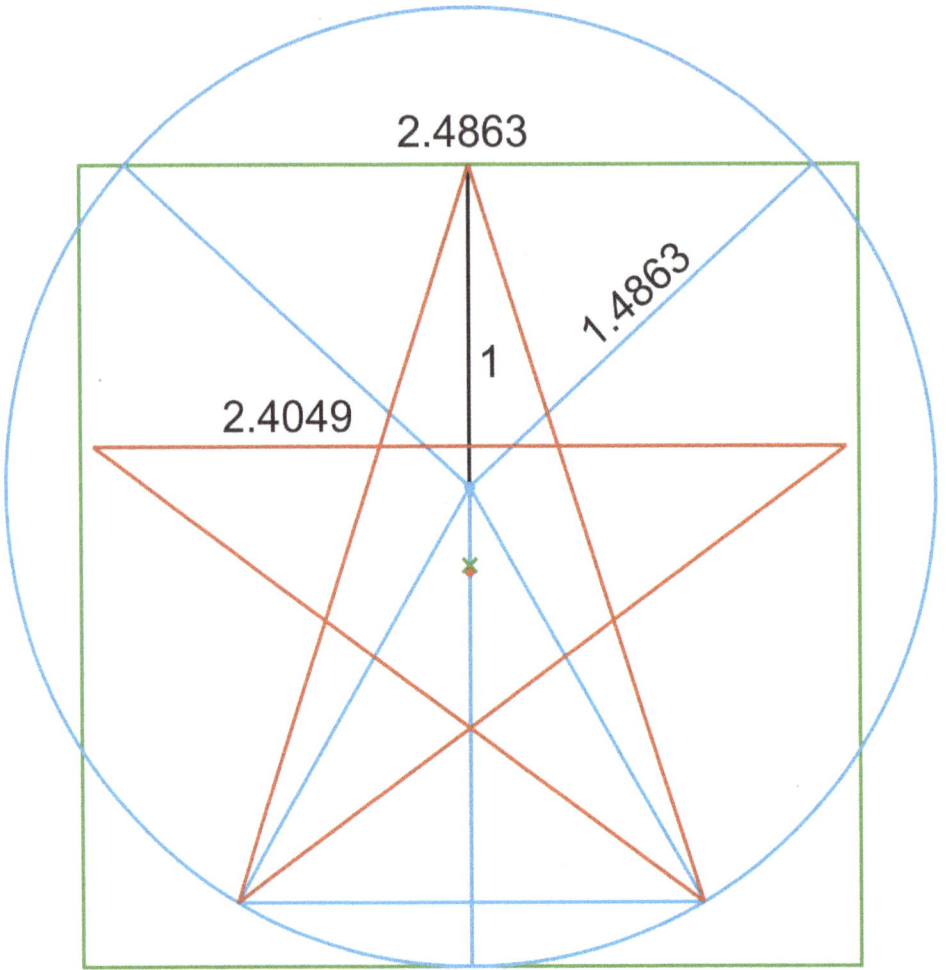

Fig. 18.1: *Model Vitruvian Martian based on Da Vinci's writings with the addition of a pentagram. The model integrates a pentagram, a circle, a square and an equilateral triangle into a single figure which represents the fundamental shape of the human body. The distance between the navel and the top of the head was set to a length of 1 arbitrary unit. The values for the radius of the circle, the length of a pentagram line and the side length of the square are shown in terms of the arbitrary unit.*

The procedures used are too complex to describe here and are the best approximations that I could accomplish with my software. The fit to Da Vinci's drawing at first glance appears to be quite reasonable. The centre of the circle fits the man's navel, and the equilateral triangle and 2 bottom star points coincide with the spread-apart pair of feet. However, there is a very noticeable discrepancy with the size of the square. Da Vinci's square is approximately 1.22% smaller than that of my model despite the circles being the same size. This causes the top of the man's head to be lower than

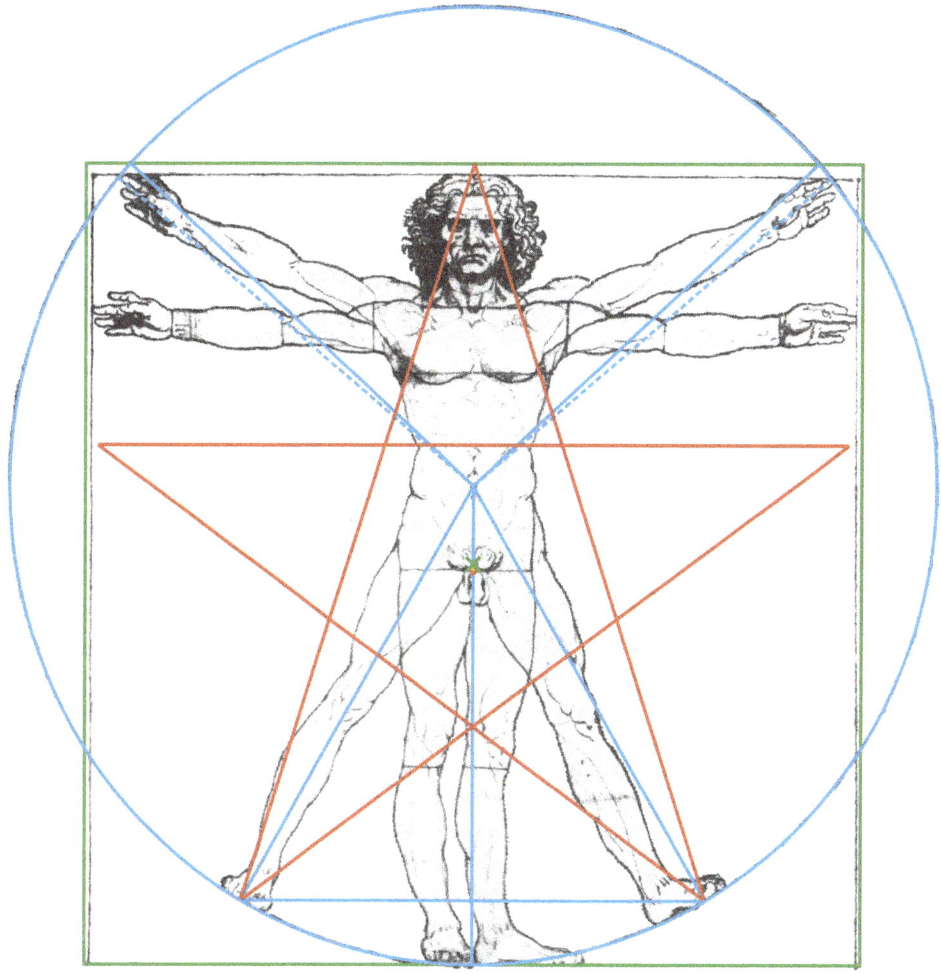

Fig. 18.2: *Overlay of the model Vitruvian Martian on Da Vinci's Vitruvian Man (McMurrich). The model was sized to overlay its circle on Da Vinci's circle. Although there is an excellent fit of the 2 pairs of feet to the model pentagram, circle, square and equilateral triangle, the remainder of the model deviates slightly from Da Vinci's drawing due to the smaller size of Da Vinci's square. The dotted blue lines to Da Vinci's upper pair of hands are 100° apart whereas the pair of solid blue model lines had to be set 95.43° apart in order to reach the intersection of the circle with the model square's upper side.*

the tip of the upper pentagram star point. It also causes the middle fingers of the upper pair of arms to touch the circle at different points than the model upper pair of radius lines. So did Da Vinci not base his model on a virtual pentagram, or did he deliberately shorten the height of the man so that the middle fingers could touch the circle at lower points? It would appear that he wanted to create the impression that the radius lines for the arms were set at an angle of 100° apart instead of the mathematically required 95.43° in my model. This would also allow each of these lines to be at an angle of 100° from the radius line for the leg on the same side of the body. With these angles, the fit of the navel-centred radii representing the upper limbs of the Vitruvian Man is almost perfect, touching the lower edge of the middle finger for the right arm and the upper edge of the middle finger for the left arm (dashed blue lines in Fig. 18.2). Another very important result of the shortening of the man's stature is that the position of the centre of the pentagram ends up being within 0.5 % of the middle point of the man's height. The centre of Da Vinci's square also occurs here, so he may have intended to have the 2 centres coincide. This suggests that Da Vinci did indeed intend to fit the pentagram to the human body but did not include it in his drawing since this might have offended religious authorities. Note that Agrippa published his pentagram drawing in a book on the occult. Da Vinci's written comments state that the beginning of the genitals (indicated by a horizontal line in the Da Vinci drawing) mark the middle of the man. Positioning the centre of the pentagram here would emphasize the importance of φ in nature and the pentagram's symbolism of fecundity and creation. It should also be mentioned that the spread apart pair of legs did not shorten the man's height by 1/14. The actual measurements show a slightly greater shortening, specifically 1/12.486 for my model and 1/12.331 for Da Vinci's Vitruvian Man.

I now wanted to fit my model to the 4 mountains, but I was faced with a dilemma. Should I fit the Tharsis Montes to my model square or to Da Vinci's square? At first it seemed that the most logical procedure to follow would be to fit the mountains to the model square since the close proximity of the Pentagram Pyramid to the 4 major mountains suggests that a fit to the pentagram would be more important to the Martian architects than the 100° angles between the radii to the limb extremities or the desire to have the pentagram centre coincide with the centre of the square. However, I finally decided that Da Vinci's square was indeed the correct one to use. What changed my mind was the amazing discovery that the northeast star point of the Pentagram Pyramid pointed to the position of the Vitruvian Man's navel when Da Vinci's top of the square was aligned to the Tharsis Montes (Fig. 18.3). However, the position of the navel first had to be located properly on the line between the rhumb model (Chapter 5) centres of Pavonis Mons and Olympus Mons by correcting for the nonlinearity of distance on the MOLA map. Simply laying the Vitruvian Man drawing on the map as was done in Fig. 14.3 creates a distorted picture since it attempts to place a 2 dimensional drawing on a spherical surface. I also slightly repositioned the longitude of the Pentagram Pyramid so that its northern star point was aimed directly at the rhumb model location of Olympus Mons rather than its

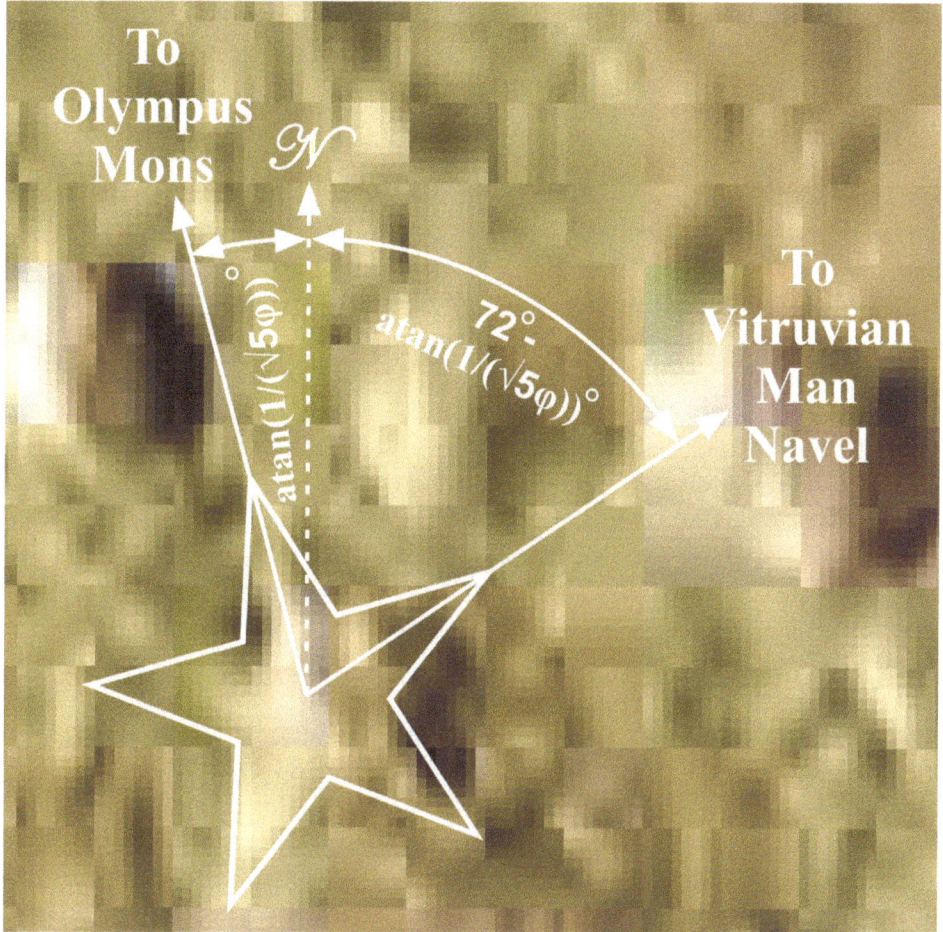

Fig. 18.3: *The Pentagram Pyramid acts as a directional compass with its 2 northern star points pointed at key locations for the construction of the Martian equivalent of the Vitruvian Man. Together with the imaginary line between the rhumb model centres of Pavonis Mons and Olympus Mons it measures out the distance and location of the navel above the bottom pair of feet. USGS Astrogeology.*

survey centre. This enabled all my measurements to be consistent with the mountain coordinates of the rhumb model from Chapter 5.

Hence, the Pentagram Pyramid placed at planetographic $\pi°$ N is located such that when the northwest star point is focused on the rhumb centre of Olympus Mons while maintaining a bearing angle of atan$(1/(\sqrt{5}\varphi))°$, the northeast star point is aimed directly at the navel of the Martian equivalent of the Vitruvian Man. With this alignment, the Pentagram Pyramid becomes a key which unlocks the essential dimensions and positions of 3 geometric shapes fitted to the Martian body, namely the circle, triangle and pentagram. The square is already defined by the 4 mountains. The navel is located by the intersection of (1), the compass line from the northeastern star point of the Pentagram Pyramid and (2), the imaginary line between the rhumb model centres of Pavonis Mons and Olympus Mons (Fig. 18.4). I calculated its coordinates to be 239.4012° E 8.2077° N. The distance between the navel and rhumb model centre of Olympus Mons is the radius of the circle. Rotating this line by 30° in the counterclockwise direction about the navel locates the position of the pentagram star point corresponding to the Martian left foot of the spread-apart pair of legs and the left vertex of the equilateral triangle. Rotating the radius line from here by 60° in the clockwise direction locates the corresponding triangle vertex and pentagram star point for the right foot. To locate the position of the middle fingertip of the right hand of the upward pair of arms, I calculated that the radius line has to be rotated by a further 100.69° in the clockwise direction rather than an even 100° in order to touch the top of the square lying between Ascraeus Mons and Arsia Mons. Then to locate the position of the middle fingertip of the left hand, it has to be further rotated in the clockwise direction by 98.62°. The construction of the equilateral triangle is completed by drawing a line equal to the length of the radius line between the vertices of the right and left feet. Next, a line equal to the length of one of the 5 lines of the pentagram can be created by multiplying the length of the circle radius line by $\varphi = 1.6180$. One end of this line can then be positioned at the right foot vertex and rotated so that the line aligns with the vertex of the left foot. Then rotating the line about the right foot vertex by 36° in the counterclockwise direction will cause the other end of the line to mark the location of the pentagram star point vertex which lies underneath the left arm (Fig. 18.5). Repeating this procedure with a pentagram line rotating clockwise by 36° about the left foot vertex from an orientation in which the line passes through the right foot vertex will mark the location of the pentagram star point vertex which lies underneath the right arm (Fig. 18.6). Rotating this line a further 36° in the clockwise direction will position the end of the line at the pentagram vertex which lies just above the head of the Martian body. Now that all 5 vertices of the pentagram have been located they can be simply joined by 5 straight lines to form the complete pentagram.

The above method for the construction of the geometric shapes fitting the Martian equivalent of the Vitruvian Man, which I will now call the Vitruvian Martian, is strictly valid only for a 2 dimensional surface. With the spherical surface of Mars, some distortion is inevitable. Depending on what distortions the architects decided to accept, the lengths and bearings of the lines and the curvature of the circle would have required alterations from their 2 dimensional values to achieve a pleasing fit to the spherical surface of Mars. However, the coordinates of the centres of the square and

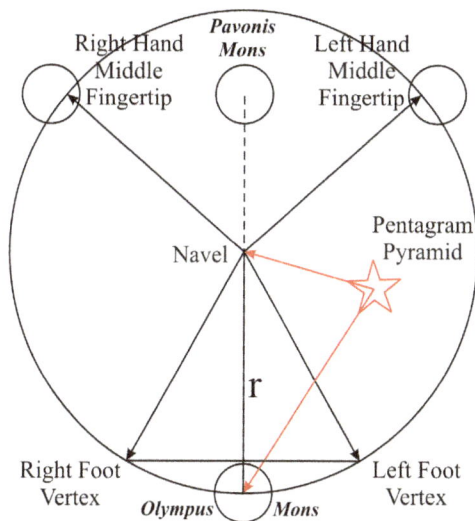

Fig. 18.4: *Construction of the fit of the Vitruvian Martian to a circle. The circle radius is determined from the intersection of the 2 northern star point compass lines from the Pentagram Pyramid with the line between Pavonis Mons and Olympus Mons. This line section is then rotated about the navel 30° counterclockwise to locate the left foot vertex, then 60°, 100.69°, and 98.62° clockwise to locate the right foot vertex and the middle fingertips of the right and left hands respectively. Placing a line equal to the circle radius between the vertices of the right and left feet completes the equilateral triangle.*

Fig. 18.5: *The left star point of the pentagram fitting the Vitruvian Martian is located by placing the end of a line equal to the radius of the circle multiplied by φ at the right foot vertex. It is first rotated about this point until it aligns with the left foot vertex, and then it is rotated 36° in the counter-clockwise direction. The opposite end of the line then marks the location of the left pentagram star point.*

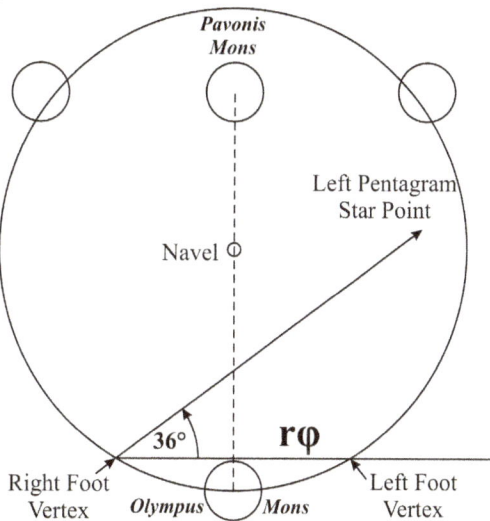

Fig. 18.6: *The right star point of the pentagram fitting the Vitruvian Martian is located by placing the end of a line equal to the radius of the circle multiplied by φ at the left foot vertex. It is first rotated about this point until it aligns with the right foot vertex, and then it is rotated 36° in the clockwise direction. The opposite end of the line then marks the location of the right pentagram star point. A further 36° rotation marks the location of the pentagram star point above the head.*

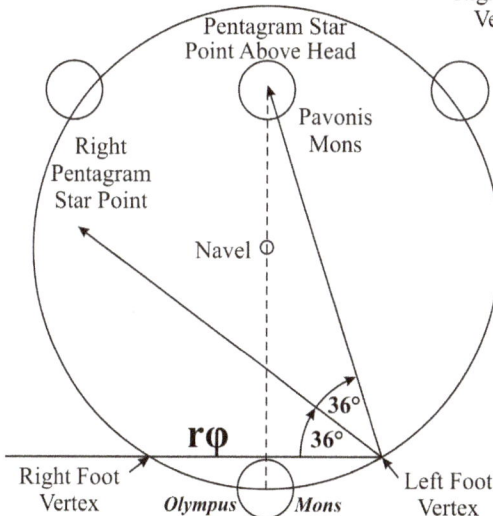

pentagram can be calculated since they lie on a line whose length is known to be $R/\sqrt{5}$ = 1518.82 km. Thus the centre of the square is at 237.3341° E 9.9738° N and the centre of the pentagram is at 237.2886° E 10.0126° N. They are only 3.51 km apart. The distance from the centre of the square to the centre of the navel is 159.99 km which is only 0.14 km more than the sacred distance of $\pi\varphi R/108$ km. The distance of 163.50 km between the centres of the pentagram and navel is only 0.10 km more than $\sqrt{3}R/36$ km. Note that 36 and 108 are the sizes of angles found in a pentagram. A picture of the square, circle, equilateral triangle and pentagram fitting the virtual Vitruvian Martian represented by the 4 giant mountains appears in Fig. 18.7. It shows a faint outline of Da Vinci's Vitruvian Man underlying the 4 geometric shapes. This picture is for illustration purposes only as it does not take into account the nonlinearity of distance in the MOLA map.

The radius of the pentagram [782.09 km, or -0.15 km less than $\sqrt{5}R/(6\varphi)$ km, measured at the star point above the head] for the Vitruvian Martian multiplied by π/φ is only about 0.02% shorter than the side of the square. Hence, the square is now almost perfectly integrated into the dimensions of the other 3 geometric shapes. I found that the distances and bearing angles of various sites to the centres of the pentagram, square and navel can be expressed in terms of meaningful sacred formulae with very small deviations from the actual measurements (Table 18.1). The 2 possible sacred distance formulae from the Pentagram Pyramid to the navel [$\sqrt{5}/(8\sqrt{3})$ km and $\pi R/(12\varphi)$ km] contain elements which refer to the circle (π), the equilateral triangle ($\sqrt{3}$), the pentagram (φ, $\sqrt{5}$) and the square (4 is a factor of both 8 and 12). The sacred distance formula from the Pentagram Pyramid is the same to the centres of the square and pentagram, namely $\pi R'/(9\sqrt{5})$ km. It does not reference the square but has elements which could be associated with the pentagram ($\sqrt{5}$, π, and 9, which is 1/4 the size of an angle of a star point). The sacred distance formula to the rhumb model centre of Olympus Mons from the Pentagram Pyramid is $\pi R'/(8\sqrt{2})$ km which references the square ($\sqrt{2}$, and 8, of which 4 is a factor) and the circle (π), both of which pass through the centre of Olympus Mons. The 2 close fitting bearing angle sacred formulae of the line from the Pentagram Pyramid to the navel [$18\pi°$ and $\mathrm{acos}(\sqrt{3}/\pi)°$] reflect the circle ($\pi$) and the equilateral triangle ($\sqrt{3}$) as well as the pentagram (18). The formula of $4\pi^2$ ° for the bearing angle to the centre of the square reflects the 4 equal sides of a square. However, the closest bearing angle formula to the pentagram centre ($16\sqrt{2}\sqrt{3}°$) reflects the square ($\sqrt{2}$, 16 = 4^2) and the equilateral triangle ($\sqrt{3}$) rather than the pentagram. Perhaps the less well fitting formula of $15\varphi^2$ = 39.2705° was intended as well. The sacred distance formulae to the centres of the geometric shapes of the Vitruvian Martian from other nearby sites associated with Biblis Tholus, Ulysses Tholus and Jovis Tholus all

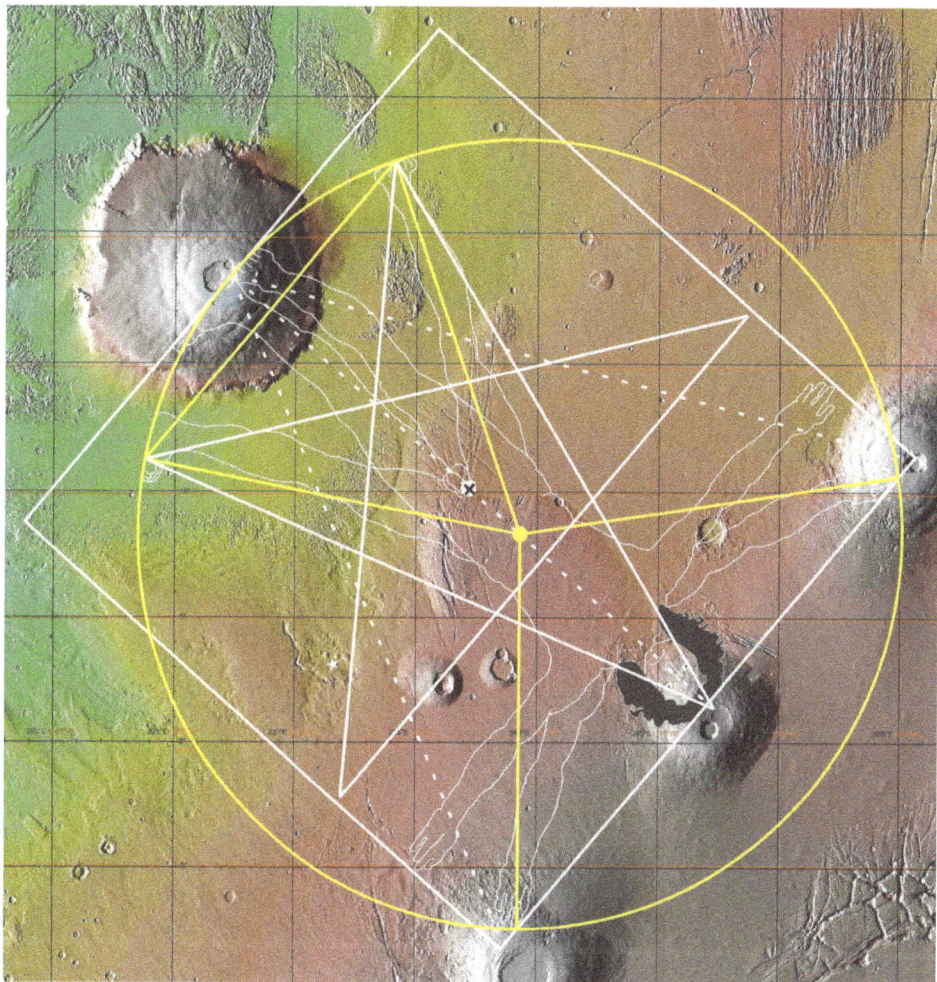

Fig. 18.7: *Vitruvian Martian model fitted to the giant mountains shows a fit of a virtual Martian body to a pentagram, square, circle and equilateral triangle. Da Vinci's Vitruvian Man is in the background. The navel is represented by a small yellow circle, and the centre of the pentagram by a small white circle. The black X on the white circle represents the centre of the square. The Pentagram Pyramid is shown as a tiny white star west of Biblis Tholus. The figure is distorted (including the locations of the centres) since it does not take into account the spherical nature of the planet, and is for illustration purposes only. The circle radius and a side of the equilateral triangle are in φ relationship to the side of a pentagram star point, and a side of the square is very close to being equal to π/φ times the radius of the pentagram. USGS Astrogeology.*

contain strong references to the circle (π), square (12 and 16 have 4 as a factor), pentagram (5, 15, 27, φ, $\sqrt{5}$) and equilateral triangle ($\sqrt{3}$). The sacred distance formula from Issedon Tholus to the tip of the pentagram star point above the head of the Vitruvian Martian is interesting since it is 3 times the

Table 18.1: Sacred Formulae to Key Vitruvian Martian Sites

A. Distance Formulae

From	To	Formula	Km	Actual	Diff.
Pentagram Pyramid	Centre of Navel	$\sqrt{5}R/(8\sqrt{3})$	548.06	548.77	0.71
Pentagram Pyramid	Centre of Navel	$\pi R/(12\varphi)$	549.51	548.77	-0.74
Pentagram Pyramid	Centre of Square	$\pi R'/(9\sqrt{5})$	527.05	527.49	0.44
Pentagram Pyramid	Centre of Pentagram	$\pi R'/(9\sqrt{5})$	527.05	527.56	0.51
Pentagram Pyramid	Olympus Mons	$\pi R'/(8\sqrt{2})$	937.50	937.98	0.47
Ulysses Tholus N Crater	Centre of Navel	$R'/12$	281.35	281.48	0.13
Ulysses Tholus	Centre of Navel	$\pi R/(15\sqrt{5})$	318.10	318.35	0.25
Biblis Tholus Caldera N	Centre of Navel	$R/(5\sqrt{3})$	392.16	392.06	-0.09
Jovis Tholus West	Centre of Square	$eR'/16$	573.59	573.59	0.00
Pavonis Mons	Centre of Square	$R/(2\sqrt{5})$	759.41	759.41	0.00
Olympus Mons	Centre of Square	$R/(2\sqrt{5})$	759.41	759.41	0.00
Biblis Tholus Caldera N	Centre of Pentagram	$\sqrt{5}\varphi R/27$	455.09	455.10	0.01
Issedon Tholus	Centre of Pentagram	$\pi\varphi R/8$	2157.94	2157.39	-0.55
Issedon Tholus	Head Star Point Tip	$3R/(2\sqrt{5})$	2278.23	2278.68	0.44

B. Bearing Angle Formulae

From	To	Formula	Degrees	Actual (Abs. Val.)	Diff.
Pentagram Pyramid	Centre of Navel	18π	56.5487	56.5496	0.0009
Pentagram Pyramid	Centre of Navel	$acos(\sqrt{3}/\pi)$	56.5418	56.5496	0.0078
Pentagram Pyramid	Centre of Square	$4\pi^2$	39.4784	39.4740	-0.0044
Pentagram Pyramid	Centre of Pentagram	$16\sqrt{2}\sqrt{3}$	39.1918	39.0893	-0.1025
Pentagram Pyramid	Centre of Pentagram	$15\varphi^2$	39.2705	39.0893	-0.1812
Pentagram Pyramid	Olympus Mons	$atan(1/(\sqrt{5}\varphi))$	15.4504	15.4503	-0.0001
Ulysses Tholus N Crater	Centre of Navel	π^2	9.8696	9.8127	-0.0569
Ulysses Tholus SE Crater	Centre of Navel	$2\sqrt{3}$	3.4641	3.4930	0.0289
Biblis Tholus Caldera N	Centre of Navel	$11\varphi^2$	28.7984	28.7951	-0.0033
Centre of Navel	Issedon Tholus	25φ	40.4508	40.4260	-0.0248
Centre of Navel	Issedon Tholus Square	$18\sqrt{5}$	40.2492	40.2424	-0.0068
Centre of Square	Jovis Tholus West	$22\sqrt{2}$	31.1127	31.0996	-0.0131
Pavonis Mons	All 3 Centres	$atan(\pi/e)$	49.1318	49.1318	0.0000
All 3 Centres	Olympus Mons	$atan(\pi/e)$	49.1318	49.1318	0.0000
Biblis Tholus Caldera N	Centre of Square	$6\sqrt{2}$	8.4853	8.4728	-0.0125
Centre of Square	Issedon Tholus	$6e^2$	44.3343	44.3435	0.0092
Ulysses Tholus SE Crater	Centre of Pentagram	$6\sqrt{5}$	13.4164	13.4269	0.0105
Ulysses Tholus Caldera	Centre of Pentagram	$4\varphi^2$	10.4721	10.4773	0.0052
Biblis Tholus Caldera N	Centre of Pentagram	5φ	8.0902	8.0958	0.0056

sacred distance between Pavonis Mons and either Ascraeus Mons or Arsia Mons. The bearing angles from the nearby sites also have sacred formulae which honour the circle (π, π^2), square (4, $\sqrt{2}$), pentagram (5, φ, φ^2, $\sqrt{5}$, e) and equilateral triangle ($\sqrt{3}$). The bearing angles from Issedon Tholus sites honour the pentagram (18, 25, $\sqrt{5}$, e^2).

In summary, the Pentagram Pyramid acts like a key from which the fit of the Martian body to the geometric shapes of a circle, equilateral triangle and pentagram can be constructed and located in the bisected isosceles triangle monument site along with the square. The shape of the key alludes to the fit to a pentagram, and its latitude ($\pi°$ N), to the fit to a circle. The radius of the circle and the length of a side of the equilateral triangle are in φ relationship to the side of a pentagram star point. The radius of the pentagram multiplied by π/φ is only about 0.02% smaller than the side length of the square. No structures marking the presence of either the geometric shapes or the Martian body itself were observed on the MOLA map, suggesting that the Vitruvian Martian has only a virtual existence. Since Da Vinci used the same relationship of the dimensions of the square to the dimensions of the circle as the Martian version of the Vitruvian Man, it would seem that his Vitruvian Man drawing was ultimately derived from the planet Mars. Da Vinci must have had access to secret documentation which was probably also available to the Roman architect Vitruvius. Whether or not he actually knew that it came from Mars can only be speculated. It is also likely that Da Vinci was aware of the fit of the body to the pentagram but refrained from including it in his drawing for political reasons. Together, the Pentagram Pyramid and the bisected isosceles triangle monument should be regarded as a single construct and probably the most important monument on Mars. It shows the tremendous regard that the Martian people must have had for their bodies. It is measured in terms of the planetary radius. It contains many of the most important angles found on Mars, i.e., 30°, 60°, 18°, 36°, 72°, 108°, and 45° (by implication as the 1/2 value of 90°). It contains all of the 6 important irrational numbers (i.e., π, φ, e, $\sqrt{5}$, $\sqrt{3}$ and $\sqrt{2}$) and the important integers of 3, 4 and 5. It represents all of the important spiritual messages, i.e., the sacredness of the Martian body, the sacredness of the planet, and the infinity of the Divine symbolized both in irrational numbers and in the pentagram which represents creation and fecundity. The protrusion of the pentagram star point above the head might even represent the 7th chakra whose purpose is to connect the person to the Divine.

Concluding Remarks

With the finding that the mountains of Mars are arranged in precise geometrical patterns which employ sacred geometry in lengths, angles and coordinate positions, it has become obvious that these colossal mountains can no longer be regarded as wonders of Nature. In the final analysis they were engineered, not only in their positioning but also in their very construction as evidenced by the existence of linear sections in their perimeters and calderas with bearing angles mainly honouring the pentagram (36°, 54°, 72°), the equilateral triangle (30°, 60°) and possibly the square or rectangle (45°). While most artifacts on Mars such as the many faces and pyramid structures are extremely difficult to date, scientists using crater counting techniques have estimated that the mountains must have originated about 3 - 4 billion years ago. This gives us a good estimate of how long intelligent life has been in contact with Mars and possibly with other bodies in the solar system such as our own planet Earth. It is a mind boggling amount of time!

Whether or not the mountains and rises also represent a means of relieving internal planetary pressure, construction sites enclosing living spaces, or refuse from the hollowing out of a portion of the planet's interior in order to create a huge biosphere has yet to be objectively determined. We have been conditioned to look for life only on planets that could possibly have a biosphere on the outside surface. It is highly plausible that most civilizations exist in subterranean biospheres that are much easier to establish, hide, defend and stabilize than exterior ones. Creation of subterranean biospheres would greatly increase the habitable space of a solar system. Although they would involve a lot of engineering, we must realize that outside surface biospheres may also require a lot of engineering in terms of establishing an atmosphere together with a compatible, complete and self-sustaining system of flora and fauna. The earth's biosphere may not have come about simply by fortuitous circumstances and evolution working over long periods of time. We must be open to the possibility that our environment too has been engineered. When one considers all the factors that had to have been put into place for our Earth to succeed as a seemingly self-developed and self-sustaining home for complex life forms such as ourselves, perhaps we will begin to suspect that in actual fact it could not possibly have come about simply by chance after all. It may still be under the unseen care of a higher intelligence and technology to keep it going! The presence of an exterior biosphere on this planet does not necessarily exclude the possibility of one or more subterranean biospheres existing alongside our own. The UFO literature is full of accounts of spacecraft entering remote mountains and plunging into bodies of water. Perhaps we share this planet with one or more non-human species who live either undetected amongst us or else in underground bases[2]. After all, earth has been around for a long time, and our biosphere has only had significant oxygen levels for approximately the last 2.4 billion years. If intelligent species were present in the solar system 3 - 4 billion years ago, and if they had established any kind of permanent presence on Earth during that era, they would probably have had to create one or more artificial subterranean biospheres.

The possibility exists that Mars could be an artificial planet in part or its entirety. We need only to look at some of the moons in our own solar system to realize that this may not be out of the reach of a highly advanced civilization. Some of these moons may in all likelihood be totally artificial – more like enormous spaceships than natural solar system bodies. Prime candidates for this are the Earth's moon, the Martian moon Phobos, and Iapetus, a moon of Saturn. The Earth's moon has seismically rung hollow after the Apollo XII lunar module crashed into it on November 20, 1969[3]. It's orbit is placed at 60 Earth radii from our planet. A circle whose radius is equal to the sum of the radii of the moon and the Earth has a circumference virtually equal to the perimeter of an imaginary square fitted about the planet. Are we to believe that this has happened just by chance within the confines of ordinary physics? Iapetus has been hypothesized by Richard C. Hoagland to be a truncated icosahedron (a polyhedron with 12 regular pentagonal faces and 20 regular hexagonal faces) rather than a true sphere[4]. It has an 800 kilometer long ridge along the equator which is an impressive 20 kilometers high, rivalled only by Olympus Mons in the solar system. Hoagland suggests that this moon must therefore be artificially constructed and, since its density is only 1.21 or slightly greater than water, hollow. Soviet astrophysicist Shklovsky and Dr. S. Fred Singer, special advisor to President Eisenhower on space developments speculated that Phobos was artificially constructed and hollow[5]. Raymond H. Wilson Jr., Chief of Applied Mathematics at NASA agreed, stating that "Phobos might be a colossal base orbiting Mars"[5]. The same might be true for Deimos.

The ultimate conclusion from the findings of this book is that we are truly not alone. If NASA and major world governments have been engaged in a cover-up regarding the existence of extraterrestrials, they will no longer be able to do so. You cannot hide the tallest mountains in the solar system. Their arrangement in artificial patterns and the meaningful bearing angles of their linear segments have now been revealed for everyone with an open mind to see. What is difficult to understand is why extraterrestrials have not identified themselves to us. Are they maintaining silence to quarantine us due to our aggressive tendencies, and are patiently waiting until we evolve to a point where we can interact more harmoniously with them? Or is there another even less palatable reason such as being prevented from doing so by some sinister element, e.g., a warrior species that may have defeated the original inhabitants of Mars and possibly now controls our own planet covertly as well as Mars?

What we should do with this new knowledge is an open question and partly depends on those who are now in control of the planet. Undoubtedly we need to rethink our own origins and history. But we must also plan for the future, a future that includes contact and interaction with other intelligent civilizations not only in this solar system, but also in the galaxy and possibly in the entire universe itself. Hopefully it will be a peaceful and mutually beneficial relationship, one in which we will be given much information about the past and about what exists beyond the limits of our telescopes.

As a start, humanity needs to engage in the pursuit of truth. Without awareness of our true situation we will not know what we have to deal with. In order to do this, we must take individual responsibility for obtaining the truth. Most of us now get our "truth" from our educational system, our religions, and the mass media, all of which are under the control of our "leaders". We are taught that official channels of information are the best we can do for solid data and that we can only trust authoritative sources. The sooner that we can break away from this mentality, the more chance we have for a healthier view of reality. To do this, we have to vastly broaden our sources of information. We have to read lots of books, search the Internet, and go to hear talks given by individuals who are independent thinkers. Be wary of those sources which argue from the point of authority and put down other sources without giving them proper consideration. Give people who have been ridiculed and discredited a fair chance to present their viewpoints and data. No one person or source has all the information, and no one person or source is necessarily completely correct. Each has a portion of the truth and it is up to you to sort out this portion from the rest. You are the one who must ultimately work it out, but you can only do so if you assemble information from all available sources, not just the officially sanctioned ones. And the sources need not just be limited to the physical dimension. There is plenty of evidence to demonstrate that psychics are able to tap into other non-physical sources of information. Even the CIA used psychics for remote viewing. We are in a situation now where truth seems false and threatening, and the current fantasy created for us by our "leaders" seems real and protective. It is going to take courage and patience before we will be able to accommodate ourselves to the new reality that we can and should uncover.

Humanity must then assess and improve its current level of ethical and spiritual development to prepare itself for interfacing with alien civilizations which are vastly more technically and spiritually advanced than our own. Otherwise it will remain vulnerable to being controlled by forces which remove freedom and dignity. Each of us must first decide how to deal with our own character flaws on an individual level. Then we must choose how we are going to react to those who abuse others by taking inappropriate power and who cater to self-serving agendas at public expense rather than empowering everyone to develop in knowledge, wisdom and creativity. If we choose violence, we are no better than the abusers, and will only perpetuate the cycle of fear, anger, hate, aggression, and destruction. This is hardly the way to prepare ourselves for citizenship in the wider galaxy/universe. No, we must instead reflect on ourselves and take this as an opportunity for us to grow somehow from within. First and foremost we must develop a caring and compassionate attitude towards each and every one of us, and towards all life and our environment, starting with our own planet. Then we must ask how we might be able to bring about change in the aggressors themselves. Those who pursue inappropriate political, economic and military power seem to fear peace, intelligence and spirituality in the human population. Throughout our history, people who manifest these qualities have been persecuted and tortured, and are often put to death either through assassination, executions, wars or genocide. The reason? - these are qualities that can take away the wrongful and inappropriate power of the aggressors. Yet they are the very things that will be of most benefit to the aggressors as well as to ourselves. If we were to develop our spirituality on a global scale, how much might we be able to achieve through prayer, energy healing and intention? Recent studies in parapsychology have found that thought really does change reality – even physical reality[6,7]. Perhaps this is what our destiny is for the next precessional constellation in the zodiac, the Age of Aquarius, which is just beginning or rapidly approaching. If we can meet the challenge, a new period of peace, stability and rewarding integration with the greater universe and the Divine Energy could ensue for our children. If not, perhaps a devastating catastrophe will be our fate.

Spirituality has been denigrated during the reign of science in recent times. Science denied all but physical reality. Surely there is more to reality than what we can detect with our senses and instruments. Spiritualist mediums can perceive the presence of departed souls living in another dimension, and often are able to receive messages of support from them. Some people who have had near-death experiences could see and hear during this event even though their brains were considered by clinical assessment to be non-functional - including people blind from birth[8]. The time is now appropriate, and the necessity urgent, to reconnect with the greater reality of the Divine. This is the true meaning of spirituality, and can only be done by direct contact of the individual with the Divine rather than through any intermediary. To achieve this, one only needs to desire union with the Divine, to forgive everyone for everything, to ask the Divine for spiritual transformation, and then quiet the mind in order to open oneself up to Its transforming power. It is a gradual process which can ultimately be accomplished with only one or two brief daily meditation(s) lasting about 5 minutes each. It does not take much effort on our part. The Divine does the real work. This will help us to achieve the necessary energy balance within ourselves so that we can become beings of love who can resolve emotions and problems in a healthy way rather than powerlessly transmit, and destructively act on, fear and anger. Love is the most powerful and fundamental energy in the universe. If enough of us can operate in its light, then perhaps the more spiritually developed extra-terrestrials and/or other-dimensional beings will also come to assist us.

May the Divine Energy be with you.
We are all in this together

References

1. *Leonardo Da Vinci The Anatomist (1452-1519). J. Playfair McMurrich. The Williams & Wilkins Company, Baltimore, 1930.*

2. *Underground Bases: A Lecture by Phil Schnieder: May 1995. Food For Thought. Alien Underground Base Near Dulce, New Mexico September, 1997. Updated July 20, 2009 by columnist David Lawrence Dewey. http://www.dldewey.com/schnidr.htm*

3. *Earth's Moon and Human Evolution. Dr. N Huntley, Ph.D. The Canadian. 2005. http://www.agoracosmopolitan.com/home/Frontpage/2008/07/18/02474.html*

4. *Moon with a View: Or, What Did Arthur Know ... and When Did He Know it? Part 5. Richard C. Hoagland. 2005. http://www.enterprisemission.com/moon5.htm*

5. *Eisenhower White House On Mars' Moon Phobos Being Artificial. http://www.rense.com/general20/eisenhowerwh.htm*

6. *Entangled Minds: Extrasensory Experiences in a Quantum Reality. Dean Radin. Simon & Schuster, 2006.*

7. *The Intention Experiment. Using your Thoughts to Change your Life and the World. Lynne McTaggart. Free Press, 2007.*

8. *Lessons From the Light: What We Can Learn from the Near-Death Experience. Kenneth Ring and Evelyn Elsaesser Valarino. Moment Point Press, Inc., 2006.*

Appendix: *Coordinates (planetocentric) of 48 Important Sites on Mars*

	Longitude	Latitude
Alba Mons	249.6749° E	39.4492° N
Albor Tholus	150.3167° E	18.7937° N
Apollinaris Mons	174.1293° E	7.3920° S
Apollinaris Mons Caldera	174.2855° E	8.5654° S
Arsia Mons	238.6754° E	8.0996° S
Arsia Mons Caldera	239.5563° E	9.3152° S
Ascraeus Mons	255.5276° E	11.2791° N
Ascraeus Mons Caldera	255.6286° E	11.1668° N
AscSC1	260.0718° E	19.2776° N
AscSC1a	264.0053° E	14.0093° N
AscSC2	249.1186° E	17.9685° N
Ayacucho Crater	267.9781° E	38.1812° N
Biblis Tholus	236.1176° E	2.6278° N
Biblis Tholus Caldera N	236.2001° E	2.4112° N
Biblis Tholus Caldera S	236.2032° E	2.3197° N
Ceraunius Tholus	262.9598° E	24.1315° N
Ceraunius Tholus Caldera	262.9032° E	23.9674° N
Elysium Mons	147.1701° E	24.4910° N
Elysium Mons Caldera	146.7813° E	24.6518° N
Fesenkov Crater	273.4705° E	21.6321° N
Hecates Tholus	150.3210° E	31.5900° N
Hecates Tholus Caldera	150.1079° E	31.7286° N
Issedon Tholus	265.2990° E	36.0037° N
Issedon Tholus Caldera	265.1574° E	36.0006° N
Issedon Tholus Square	265.2990° E	36.1638° N
Jovis Tholus East	242.6544° E	18.2340° N
Jovis Tholus West	242.4932° E	18.2598° N
Nicholson Crater	195.5360° E	0.2069° N
Olympus Mons	227.3258° E	18.3570° N
Olympus Mons Central Caldera	226.7242° E	18.3211° N
Olympus Mons NE Caldera	227.1980° E	18.7525° N
Paros Crater	261.8686° E	21.9949° N
Pavonis Mons	247.1037° E	1.5993° N
Pavonis Mons Caldera	247.1724° E	0.5129° N
Pentagram Pyramid	231.6302° E	3.1045° N
Pettit Crater	186.1673° E	12.2701° N
Poynting Crater	247.2630° E	8.4113° N
Tharsis Tholus North	268.7222° E	13.7122° N
Tharsis Tholus South	269.6923° E	12.5955° N
Tharsis Tholus Caldera	269.1778° E	13.3574° N
Ulysses Tholus	238.4356° E	2.9244° N
Ulysses Tholus Caldera	238.6052° E	2.9430° N
Ulysses Tholus N Crater	238.5874° E	3.5284° N
Ulysses Tholus SE Crater	239.0599° E	2.6432° N
Uranius Mons	267.0163° E	26.0069° N
Uranius Tholus	262.4093° E	26.2088° N
Uranius Tholus Caldera	262.5647° E	26.1718° N
Uranius Tholus Round Top	262.4833° E	26.1748° N

www.ingramcontent.com/pod-product-compliance
Lightning Source LLC
Chambersburg PA
CBHW042310210326
41598CB00041B/7336